JAMES PARKINSON
HIS LIFE AND TIMES

HISTORY OF NEUROSCIENCE

Series Editors

Louise Marshall
Brain Research Institute
University of California
Los Angeles

F. Clifford Rose
Charing Cross & Westminster
Medical School
University of London

JAMES PARKINSON
HIS LIFE AND TIMES

By
A.D. Morris

F. Clifford Rose
Editor

Birkhäuser
Boston Basel Berlin

F. Clifford Rose
University of London
Charing Cross & Westminster
Medical School
The Reynolds Building
London W6 8RP, England

Library of Congress Cataloging-in-Publication Data
Morris, Arthur D.
James Parkinson : his life and times / A.D. Morris ; edited by F. Clifford Rose with a foreword by John Thackray.
p. cm. — (History of neuroscience)
Bibliography: p.
ISBN 0-8176-3401-0 (alk. paper)
1. Parkinson, James, 1755–1824. 2. Neurologists—Great Britain—Biography. I. Title. II. Series: History of neuroscience (Boston, Mass.)
[DNLM: 1. Parkinson, James, 1755–1824. 2. Neurology—biography. WZ 100 P245M]
RC339.52.P376M67 1989
616.8′0092′4—dc 19
[B]
DNLM/DLC 88-36582

Printed on acid-free paper

ISBN 0-8176-3401-0
ISBN 3-7643-3401-0

Typeset by Publishers Service, Bozeman, Montana.
Printed and bound by Edwards Brothers, Inc., Ann Arbor, Michigan.
Printed in the U.S.A.
9 8 7 6 5 4 3 2 1

Preface

Dr. A. D. Morris had a long interest in, and great familiarity with, the life and times of James Parkinson (1755–1824). He was an avid collector of material related to Parkinson, some of which he communicated to medical and historical groups, and which he also incorporated into publications, especially his admirable work, *The Hoxton Madhouses.*

When Dr. Morris died, in 1980, he left behind a large typescript devoted to Parkinson's life. It was single-minded in its dedication to primary texts, quoting liberally from the whole range of Parkinson's writings. This was particularly valuable since so many of Parkinson's publications were tracts, pamphlets, or occasional pieces which are now very scarce. A copy of the entire manuscript has been deposited in the Library of the Wellcome Institute for the History of Medicine in London, where it may be consulted. The length of the manuscript made publication of the whole impossible, especially since it would have had to include the facsimile reproduction of Morris's *The Hoxton Madhouses.*

The process of editing has been far more complicated and taken much longer to achieve than was originally envisaged due to checking and shortening, but much of the flavour of Dr. Morris's manuscript has been preserved. A number of individuals have given freely of their time and expertise. Dr. W. A. Smeaton of University College, London, refashioned the chemical chapter and John Thackray of the Geological Museum showed Parkinson as a palaeontologist while also correcting the manuscript as a whole.

Dr. Janet Brown helped edit several chapters, and Ben Barkow worked throughout in editing, checking quotations, and compiling the bibliography. Several people, including Heather Edwards, have contributed to the typing.

Mrs. Dorothy Morris devotedly helped her husband over many years in his researches and typed many of the earlier versions. She graciously provided financial support while Laurence Pedersen worked for two years on the social, political, and scientific backgrounds of the times of James

Parkinson. We hope that this resulting book will please her and repay, in some measure, her infinite patience. It is also a particular pleasure to thank Dr. W.H. Bynum, for without his persistent work and encouragement, this book would not have been published. All the good editing is due to him and his team at the Wellcome Institute for the History of Medicine—all the faults are mine.

F. Clifford Rose

A Memoir of A.D. Morris

DR. JOHN THACKRAY

Arthur Daniel Morris was born on the 11th of May, 1889, in the village of Treorchy in the Upper Rhondda Valley, South Wales. His father, Dr. William Morris, was a celebrated nonconformist preacher and bard. Arthur, one of six children, went first to Porth County School, and then to Taunton School as a boarder.

He spent his school holidays with the family of Dr. John Armstrong, from whom he learnt some of the rudiments of general practice, which turned his thoughts toward a career in medicine.

Arthur took the London matriculation examination at the age of 18, and enrolled as a medical student at the University College of South Wales and Monmouthshire in Cardiff, where he studied for three years before moving to University College Hospital for his final year. He qualified LRCP and MRCS from the Hospital in October 1915 and was accordingly put on the medical register.

Arthur returned to South Wales for his first job, where he worked with Fergus Armstrong, John's younger brother, for five months. He then returned to London as Assistant Medical Officer at Hammersmith Infirmary. When the hospital was taken over by the War Office in 1916, Arthur joined the Army and continued to serve at Hammersmith as a first lieutenant until July, when he was ordered to Mesopotamia with the Royal Army Medical Corps. His war service as a Captain Surgical Specialist included time in hospitals in Bombay and Calcutta, service on the hospital ship Devonshire, and finally a spell in the Agnes Hunt and Robert Jones Hospital, Oswestry.

He obtained his M.D. from the University of Brussels in 1920 and his M.B. and B.S. from London in 1924, both while he was Assistant Medical Officer at Kensington Infirmary (later St. Mary Abbotts Hospital) under Remington Hobbs. This was followed by a second spell at Hammersmith Hospital as Deputy Medical Superintendent from 1926 to 1934 under Sir Thomas Carey Evans.

In 1935, Arthur was appointed Medical Superintendent of St. Leonard's and St. Matthew's hospitals in Shoreditch. Those hospitals presented a sorry contrast to the splendid facilities enjoyed at the Hammersmith; buildings, equipment, accommodations, and staff morale were all poor, and the first few years were not easy ones. In those days, the Medical Superintendent was responsible for every case in the hospital, although he delegated care and treatment through his deputies. He could be no narrow specialist, as he had charge of medical, surgical, gynaecological, and children's beds, as well as maternity and mental wards. Arthur took a particular interest in orthopaedic work and started a special finger clinic to deal with accidents in the nearby furniture factories. He worked heroically beside his staff and the people of Shoreditch through the war, and his changed attitude to the place is shown by his refusal, in 1941, to move when a better hospital was offered to him.

In 1939, Arthur was President of the Harveian Society, an old and distinguished medical society in London named for William Harvey, the discoverer of the circulation of the blood. He read an address in which he described the work of a Medical Superintendent.

Arthur first became interested in James Parkinson in 1952. The fact that this famous medical man had lived in the borough of Shoreditch, and had been parish doctor to the workhouse which stood on the site of St. Leonard's Hospital, fascinated Arthur. He read all he could on Parkinson and on the history of Shoreditch, including the whole of Parkinson's medical and political meetings, and took up the study of geology in order to be able to understand Parkinson's *Organic Remains of a Former World*. The idea of a biography took shape in his mind, and he corresponded widely to follow up the different facets of Parkinson's life. He was lucky to make contact with a number of Parkinson's descendants, including Miss Florence Brown, his great-granddaughter, then aged 95. Arthur was by now an enthusiast for geology books, initially concentrating on those cited by Parkinson.

He retired from St. Leonard's and St. Matthew's Hospitals in 1955, and in that year realized a long-cherished hope when he unveiled a plaque to James Parkinson in the church of St. Leonard's, Shoreditch. In February 1957 he married Dorothy A. Thomson in Vernon Baptist Church, near King's Cross. He and his wife lived in Peterborough until 1960, when they moved to Eastbourne.

His retirement was a long and happy one. His book on Parkinson progressed steadily for a number of years. Many chapters were completed but others remained only drafted, and in the end the task of completion proved too much. His interest in geology led him to study and write about a number of other prominent medical geologists, including Joseph Townsend, Samuel Dale, Gideon Mantell, and Sir John Hill. He did not neglect purely medical topics, and published an account of his medical friends and mentors from Treorchy, the brothers Armstrong, as well as a history of the Medical Superintendents' Society (for doctors who administrated community hospitals). He was always delighted to correspond about old books and their authors and to share his extraordinary knowledge of the East End of London and its inhabitants.

Arthur Morris died on August 15, 1980, after a long illness, during which he was devotedly nursed by his wife Dorothy.

List of Relevant Historical Papers

(1955). James Parkinson. *The Lancet 9th April 1955*:761–763.

(1957). *Hoxton Square and the Hoxton Academies*. T. Davis, London, p. 12.

(1958). *The Hoxton Madhouses*. Goodwin Bros., Cambridge, p. 25.

(1960). *'Sir' John Hill* M.A., M.D. (1706–1775) *apothecary, botanist, playwright, actor, novelist, journalist*. Proceedings of the Royal Society of Medicine *53*:55–60.

Contents

Preface v

A Memoir of A.D. Morris by Dr. John Thackray vii

CHAPTER 1 London in 1800 1

CHAPTER 2 James Parkinson: A Biographical Overview 10

CHAPTER 3 The Political Radical 23

CHAPTER 4 The Pop-Gun Plot 38

CHAPTER 5 The Popular Medical Writer 54

CHAPTER 6 Medical Literature 62

CHAPTER 7 Medical and Surgical Cases 68

CHAPTER 8 The Parish Doctor 77

CHAPTER 9 James Parkinson's *Chemical Pocket Book* 88

CHAPTER 10 The Madhouse Doctor 96

CHAPTER 11 Mr. Parkinson, the Palaeontologist 114

CHAPTER 12 A Discussion of Parkinson's *Essay on the Shaking Palsy* 131

Appendix
An Essay on the Shaking Palsy (original text) 151

Parkinsonism Since Parkinson by F. Clifford Rose 176

References 188

Index 199

CHAPTER 1

London in 1800

The place where James Parkinson was born, Hoxton, was little more than a village, half a mile from one of the northern gates of the City of London. In the latter half of the 18th century it had two main thoroughfares, Kingsland Road and Old Street Road. Kingsland Road is the continuation of Bishopsgate, both part of an old Roman road which continued northward to Lincoln and York. Old Street Road (now Old Street) joins Kingsland Road at a right angle opposite the parish church of St. Leonard, and in this angle lies Hoxton. Hoxton Old Town (now Hoxton Street) began at Old Street and ran parallel to Kingsland Road. Interlacing these main thoroughfares there were small streets, courts, and alleys, but beyond were open fields, market gardens, flower gardens, orchards, and grand old mansions standing in extensive grounds. Dating from the 17th century were two most fashionable residential squares. One of these was Hoxton Square, and it was at No. 1 that James Parkinson was born, lived, and practised his profession.

Hoxton appears in the Domesday Book as Hocheston, but Hogsdon, Hogsden, and Hoxdon were variations through the centuries. The first church dedicated to St. Leonard was built A.D. 900. A second church on the same site lasted for centuries until, on Sunday December 30th, 1736, whilst a service was in progress, worshipers were startled by a loud rending, cracking noise and saw the walls and roof starting to collapse. Panic stricken, the congregation fled to the door and, in the crush and commotion that ensued, many were injured. The edifice was condemned as unfit for further use and had to be demolished. The building of the third and present church commenced in 1736 and was completed in 1740. It is a fine example of Georgian ecclesiastical architecture, the work of George Dance, the elder (1700–1768), favorite pupil of Sir Christopher Wren (1632–1723) near whom he lies buried in St. Paul's Cathedral. It was the first church to be lit by gas, "an innovation most profane" (Bradley, 1914: 12). James Parkinson was baptized and married in this church, and when he died was buried within its precincts. It has long been known as the Actors Church because a number of Elizabethan actors, most of them contemporaries of Shakespeare, were buried there. A marble plaque, affixed to the north wall in the interior of the church, commemorates "the work done for the English Drama by the players, musicians, and other men of the theatre who are buried within the precincts of this church and in particular those who are named below."

The first English Playhouse was built by James Burbage in Shoreditch in 1576. Prior to this time play actors had given performances in the city inns whose balconies formed natural galleries for the watching of entertainments in their courtyards. Burbage's venture proved so successful that in 1594 he built a second playhouse, near the first, which he called The Curtain. The theatre had an innovation, a stage, but The Curtain also had a green front curtain, hence the name. The grasping ground landlord became troublesome when the lease was expiring and refused to renew it. Called simply The Theatre, it was a wooden structure, and in order to prevent it becoming the property of the landlord, it was pulled down. The timber was then transported to the other side of the river, where it was reerected and named The Globe Theatre (Harrison, 1939:78). It was only natural, therefore, that when Shakespeare conceived an ambition to act, he should take lodgings in Shoreditch. He obtained some employment at the Shoreditch theatres, either as a minder of the patrons' horses or as a call boy. He had not much of a reputation as an actor and never progressed beyond the playing of minor parts. When he took to playwriting, some of his plays were first performed at The Curtain, notably *Henry V* and *Romeo and Juliet*.

Shakespeare did not stay long in Shoreditch and moved to Bishopsgate. Being a meek, quiet man, and some of his fellow actors being a rowdy, hot-headed lot involved all too frequently in tavern brawls, he wanted to avoid their company away from the theatre itself. He also found that the ale brewed in Shoreditch was too strong and he had no head for it (Pearson, 1942).

In Parkinson's lifetime, the advent of a philanthropic and humanitarian spirit saw the establishment of many hospitals. In the vicinity, though not in Shoreditch itself, was established The London Hospital, where he became a medical student (Clark-Kennedy, 1962–63).

The Great Plague of 1665, which caused the death of more than 100,000 persons in the City of London, spread beyond the city walls. Shoreditch did not escape the plague's visitation, but the Great Fire which followed the plague did not reach Shoreditch.

Hoxton achieved an unwanted notoriety in the 17th and 18th centuries as a retreat for lunatics. "This hamlet has for many years acquired a melancholy distinction as the retreat of the insane and the city poor. There are three private establishments of considerable magnitude and respectability devoted to the former." Andrew Marvell (1621–1678) knew its reputation when he wrote "he is fit for nothing but Bedlam or Hogsdon." However, whereas Bedlam was a charitable institution, Hoxton's establishments were for paying patients only. The clowns at Bartholomew Fair could always provoke mirth by the mention of Hoxton.

The Hoxton madhouses were within a few hundred yards of each other, and Parkinson was the regular medical attendant to one of them, Holly House, for many years.

It was an age noted for its unrestricted callous brutality in the treatment of lunatics, criminals, and paupers. For the miserable lunatics, however, the days of chains, manacles, whippings, and dungeons were ending. The public conscience was stirred by the altruistic labors and persistent efforts of John Howard (1726–1790) on behalf of the prisoners, who lived under the most appalling conditions if they were too poor to contribute toward their upkeep. Reforms were being carried out, though somewhat tardily, which ameliorated the inhuman hardships suffered by the imprisoned. Purpose-built workhouses were being designed in which the paupers were treated humanely as victims of misfortune. Parkinson, as a member of the parish council, as a guardian of the poor, and afterwards as a parish doctor, struggled hard for social reforms, especially where the treatment of pauper children was concerned.

The prevailing vice of the age was drunkenness and this evil habit was well catered to by public houses in the proportion of 1 to every 200 inhabitants. The consumption of ale had been largely replaced by that of spirits, especially gin, which was cheap and plentiful, but the Government, moved by public opinion, did step in and impose a heavy tax on spirits which considerably reduced consumption. Parkinson, in his books on domestic medicine, warned his readers against the evil of drunkenness and condemned those who "spend a shilling on food and two shillings in the alehouse."

The parish workhouse and infirmary was built by the Trustees of the poor in the year 1777 to accommodate some 500 persons. However, during a great trade depression in the early part of the 19th century, over 1500 persons had to be accommodated. A Sunday School was established in the parish church of St. Leonard's where thousands of children were taught to read and, in Parkinson's words, also received "the inculcation of moral and religious principles."

The formative period of Parkinson's life was passed in an age full of interest for social and political historians, and students of art, literature, and architecture. As a young man, Parkinson would have followed the phases of the American Revolution, and in his more mature years the French Revolution and the long war with France and Napoleon. It was an age of aristocracy, of men and women of rank and fashion, fine clothes, and refined manners. The comedies of Richard Brinsley Sheridan have familiarized those who have witnessed their performances with the attractive and colorful costumes worn by the middle- and upper-classes of the period. For the men, there were wigs and three-cornered hats, multicoloured velvet jackets, gaudily embroidered waistcoats, knee breeches,

Figure 1. The first workhouse in the parish of St. Leonard's, Shoreditch. Built in 1777, demolished in 1863.

and silk stockings. Parkinson would have undoubtedly followed the fashion of wearing a wig, but his clothes would have been less flamboyant, of more sombre hues. Some idea of his dress can be gathered from a study of the oil paintings and engravings of residents of Hoxton Square, fellow members of the London medical societies, and his companions of the reform societies, since there is no portrait of Parkinson himself.

Parkinson's lifetime corresponded precisely with the period of the so-called Industrial Revolution which, some scholars claim, began with the ascent to the throne of George III in 1760 and ended at his death in 1820 (Mathias, 1983). At that time, James Watt produced his steam engine and Josiah Wedgwood manufactured pottery for the masses. The mining of coal and its use in the smelting of iron was developed. Transport was facilitated by the construction of new roads and the remaking of old ones. The building of bridges and docks and the creation of canals linking the large towns and seaports with the industrial areas proceeded rapidly. The laboring poor benefited from the introduction of washable cotton materials and by the gradual replacement of pewter utensils by easily cleaned crockery.

The development of agriculture did not lag behind the development of industry. The population was thought to be increasing rapidly, far too rapidly in the opinion of political economists such as Malthus, who predicted that since Britain had to be self-supporting in the matter of food, the day was not far distant when the population would not be able to feed itself and would face starvation. Parkinson thought otherwise and said that "if the population exceeded the means of support, the fault lay not in nature, but in the ability of politicians to discover some latent defect in the laws respecting the division and appropriation of property." The unprecedented growth of the population was due mainly to a decline in the mortality rate, especially that of infants and children, a factor influenced by advances in preventive medicine as well as the better skill and knowledge of the man-midwives, apothecaries, surgeons, and physicians of the time (Loudon, 1986). Side by side with the increasing population came improvements in the cultivation of arable land, resulting in bigger and better crops. Improved methods of breeding and rearing fatstock brought more and heavier pigs, sheep, and bullocks into the butchers' shops.

The City of London was more prosperous than ever. There was a phenomenal rise in the trade of the docks and in City business generally; this was reflected in Hoxton, whose population grew rapidly. The rise was due mainly to increased employment but partly because residential areas in the City were being taken over for business purposes. Houses were being demolished to make room for factories, warehouses, and offices, which forced the displaced residents to find homes beyond the City walls. The City fathers, wealthy merchants and businessmen who could afford horses and carriages, were able to live where they chose in delightful hamlets, not yet entangled by the "octopus," the ever-extending metropolis of London. The uprooted working-class families, on the other hand, were forced to seek accommodation within easy walking distance of their places of work in the City, and thus began that exodus into the Liberties beyond, including Hoxton. The open fields, gardens, orchards, and fine old houses standing in large grounds, for which Hoxton was noted, were fast disappearing and it was but a matter of time before every space was obliterated. With overcrowding and war came soaring rents. No longer could a family possess its own home but was obliged to share it with other families, until the little houses were grossly overcrowded, a condition which persisted up to the beginning of World War II. In 1788, the total number of inhabitants of Shoreditch was about 20,000, and they lived in 2502 houses. At the beginning of the 19th century, the number had increased to 34,766 persons crowded into 5732 houses (according to the first census of 1801). The total area of the parish was 645 acres, approximately a square mile, which later became the most densely populated square mile in the country. Parkinson's practice, which he carried on single-handed until 1808 when his son joined him, expanded with the increase of the population. What had once been a middle-class practice was now becoming a working-class one.

In the middle of the 19th century, overcrowding in Hoxton was at its worst. By 1851, the population was 109,209 and the number of houses 15,433, so that Hoxton was slowly becoming a poverty-stricken slum district. Prosperous residents moved into more salubrious surroundings in the ever-extending suburbs, though they continued to conduct their business in the neighbourhood. Soon, no middle-class residential streets were to be found in the parish. The majority of the laboring class, in Parkinson's time, were brickmakers, tilemakers, and clay diggers employed in the many brickworks and clay pits of Haggerston and Hackney. Some, however, were weavers employed in the Spitalfields silk industry, and others were craftsmen or laborers employed in building houses, docks, and canals.

Among the laborers were a large number of Irish immigrants, of whom Parkinson wrote: "the wretched and unprotected Irish poor, suffering . . . all the privations and miseries of exile." Poor relief was denied them and, if sick or otherwise unemployed, they faced starvation. During the rebuilding of St. Leonard's Church in 1736, an employer reduced the wages of his laboring men by 50 percent and, when the workmen refused to work for such a low wage, he imported Irishmen to do the job. This led to a clash between the local inhabitants and the Irish. A mob of 5000 attacked public houses known to be the resort of Irishmen, and a sanguinary battle took place between the English and the Irish, in which, though only one person was killed, many were injured on both sides. The local posse of constables, headboroughs, and watchmen were powerless and the militia had to be called in to quell the disturbances.

At the end of the Napoleonic war, conditions were even worse. The manufacture of munitions had ceased and some 300,000 men had been demobilised from the Army and Navy. They wandered the country, vainly searching for employment. Trade with Europe was at a standstill. Taken altogether, it was a state of affairs with which the country was to become all too familiar. The workhouse in Hoxton was grossly overcrowded, and in the years 1814 and 1815 the maintenance of the poor was costing the parish over 20,000 pounds per annum.

Sanitary reform had not yet arrived in Hoxton, and it is difficult to depict the insanitary conditions which prevailed. Only the two main thoroughfares were paved and lighted. Each house had a privy and cesspool but these were emptied only at infrequent intervals—there were only four scavengers for the whole of the parish, but any hole or corner was a privy. In the centre of the narrow streets ran a ditch, and along its length around was heaped refuse of every description, from which arose a perpetual offensive odor. Sewerage eventually found its way into the heavily polluted Thames by way of the Wallbrook, since Hoxton at this time could not boast a single sewer.

Footpads infested the dark streets at night, where they menaced the wayfarers. It was said that no decently dressed, respectable-looking person dare venture out in Hoxton at night. Parkinson, well-known in the parish as a medical practitioner, would have been exempt, as doctors always were, from the attentions of thieves and pickpockets whilst on his rounds and night calls.

It was also an age of books, tracts, and pamphlets. Printers, bookbinders, and engravers of plates abounded. Printing had become an art. The wealthy were establishing fine libraries in their spacious residences, a library being an essential part of every house if the occupant could afford it. Books appeared on every conceivable subject and the novel was beginning to make headway. Parkinson was thrilled by the novels of Mrs. Racliffe (1764–1823). Book clubs and circulating libraries sprang up in London and the large provincial towns to meet the needs of those who could not afford their own collections. Parkinson was an avid reader of books on natural history and professional subjects, as the multitude of references in his writings indicate. The numerous poetical quotations which appear in his books on domestic medicine show an innate love of poetry.

The pursuit of culture and a thirst for literary and scientific knowledge illustrated the desire of the educated classes for intellectual attainment. Many literary and scientific societies were founded in the metropolis at the end of the 18th or beginning of the 19th century, among them the Linnaean, Zoological, Astronomical, and Geological Societies, the Royal Institution, the Royal Society of Arts, the Medical Society of London, and the Medical and Chirurgical Society. Scientific periodicals began to appear; William Nicholson's *Journal of Natural Philosophy, Chemistry and the Arts,*

Dr. Alexander Tilloch's *Philosophical Magazine*, and Dr. Thomson's *Annals of Philosophy*, to each of which Parkinson contributed articles on geology. Learned societies such as the Geological, the Medical Society of London, and the Medical and Chirurgical Society commenced printed transactions in which articles by Parkinson appeared from time to time.

Hoxton some 50 years after Parkinson's death had lost its former charm, since land was completely built over. In the winter months, a foggy haze formed a perpetual canopy which the sun never penetrated. Just as, at the beginning of the century, the prosperous city folk had moved out to live in more salubrious surroundings, so now did the businessmen of Hoxton, though they continued to attend to their businesses in the parish in the day time. The population of Shoreditch eventually reached 120,000, and about half of the inhabitants lived in Hoxton.

By the end of the century, Hoxton contained some of the worst slums in London. The inhabitants, whom Parkinson would have described as the "honest poor," were poor but no longer honest, if Charles Booth is to be believed. Besant said that there was no "habitual criminal class" in Hoxton. Booth, however, in a social statistical survey of "Life and Labour in London," gives Hoxton a bad name. He says that "Hoxton is the leading criminal quarter of London, and indeed of all England. 'Wall off Hoxton,' it is said, 'and nine tenths of the criminals of London would be walled off.'" Booth, however, does not give facts and figures to back up this sweeping assertion, and somewhat condescendingly adds: "The position is doubtless exaggerated" (Booth, 1902, 3rd Series, Vol. II, 21).

The appearance of Hoxton is rapidly changing. During World War II, vast areas were laid waste by bombs, "doodle bugs," and rockets and whole streets were swept away under a scheme for slum clearance. Blocks of flats rose like the mythical phoenix from the ashes of the old Hoxton, and spread their wings higher and ever higher, ten, twelve, and more storys, into the air above. Green lawns interspersed with flower beds surrounded them on all sides, and neat, trim pathways led up to their doors, so that wide open spaces separated each block from the next and the main thoroughfares spread out into pleasant boulevards. Parks, recreation grounds, and tennis courts signify that never again will the narrow courts and alleys of Parkinson's day darken Hoxton. Where once more than 120,000 persons lived, incredibly overcrowded into one square mile (and indeed as late as 1931, some 97,000 lived in the borough), in future no more than 45,000 will reside there, housed in modern flats with modern domestic amenities. Factories, warehouses, and business premises are restricted to certain nonresidential areas and no longer permitted to encroach upon the backyards and gardens of the little houses; they have been removed and housed in purpose-built units.

St. Leonard's Church, though damaged, has now been restored, and apart from the loss of a priceless stained glass window, shows little signs of the effects of high explosives. The churchyard of St. Leonard's has been converted into a garden, the Fairchild Garden, since the beginning of this century, and its tombstones, which probably included James Parkinson's, cast away.

The 17th century houses which were to be found in Charles Square were, alas, doomed for destruction. The eastern side of the square went first, replaced by a four-story block of flats named "Parkinson Court" in honor of James Parkinson. Only the lovely old Court House in the square escaped the fate of the other houses. Hoxton Square did not escape, as many houses on the western side, including St. Peter's Church, were destroyed.

It is only within the last few years that the importance of the historical associations of Shoreditch have been appreciated, but enough has been given of the background to the Hoxton in which James Parkinson lived and labored all his life to provide some conception of its people whose lives and health lay in his hands, and others, for whose betterment he struggled so hard, so long and—it is now seen—so successfully.

CHAPTER 2

James Parkinson: A Biographical Overview

James Parkinson was born on April 11th, 1755, at No. 1 Hoxton Square in the parish of St. Leonard's, Shoreditch, in the Hundred of Ossulstone in the county of Middlesex. The baptismal register of the parish church for the month of April 1755 contains the following entry:

> James son of John and Mary Parkinson. Hoxton Square, Born 11th. Baptised 29th inst.

His father, John Parkinson, was a family doctor who had practised in Hoxton for many years; he was examined and received the Grand Diploma of the Corporation of Surgeons of London on February 7th, 1765, at the age of 39. During the years 1775 and 1776, he acted as Anatomical Warden of the Company at Surgeon's Hall. His tombstone can still be seen on the north wall of the verger's house in the churchyard of St. Leonard's, and though the epitaph is no longer decipherable, this inscription is just legible:

> To the memory of Jno. Parkinson, Surgeon late of this Parish who died on the 10th day of Jan. 1784, in the 59th year of his age.

John and Mary Parkinson had two other children: William, who was born in 1761 and died at 21 years of age, and Mary Sedgwick, who was born in 1763 and was to marry James Parkinson's close friend, John Keys.

During his apprenticeship as a surgeon to his father, James Parkinson married Mary Dale in 1781, also in the parish church of St. Leonard. Six children were born of the marriage. The first, James John, born in 1783, died in infancy and the fourth, Jane Dale, born in 1799, died at the age of 3 years. Of the surviving children, John William Keys Parkinson was born in 1785; the third, Emma Rooke, was born in 1788; the fifth, Henry William, in 1791; and the sixth, Mary Dale, in 1794.

John William Keys Parkinson was apprenticed to his father on the 7th of May, 1802, and, having been bound for six years, was granted the diploma of membership of the Royal College of Surgeons on May 6, 1808. He

Figure 2. St. Leonard's Church where James Parkinson was baptized, married and buried.

joined his father in his medical practice, the partnership of Messrs. Parkinson & Son, and after James Parkinson's death in 1824, carried on the practice for 12 years, until he moved from Hoxton Square. J.W.K. Parkinson married Diana Chapple, also in St. Leonard's Church, in 1811, and they had a daughter, Mary Hester, and a son, James Keys, who was in his turn bound apprentice to his father, becoming Licentiate of the Society of Apothecaries in 1834. He joined his father in the Hoxton practice, but after the death of J.W.K. Parkinson in 1838, moved to Milverton in Somerset, where his father-in-law was curate. He later returned to Hoxton with his family and lived at No. 1 York Place, on the Kingsland Road, until his death in 1849. He was buried in the churchyard of St. Leonard's with his

Figure 3. 3 Pleasant Row, Kingsland Road.

great-grandfather, grandfather, and father. Thus, four generations of the Parkinson family were surgeon-apothecaries in Hoxton, the firm of Messrs. Parkinson & Son having practised for a continuous period of more than 80 years.

James Parkinson died on December 21, 1824 at No. 3 Pleasant Row, Kingsland Road, Hoxton. Since he was the parish doctor, his son notified the Board of Trustees of the Poor of the Parish of St. Leonard's of his father's death. As entered in the minutes of the meeting of the Board held on December 23, 1824, his letter reads:

Hoxton Square
Dec. 23rd 1824

Gentlemen,

With feelings of deep regret I have to announce to you the death of my much beloved and respected Father.

He was suddenly attacked on Sunday afternoon with a severe paralytic affection which deprived him of his speech and of the use of his right side. Every means were employed for his recovery, but they proved to no avail—and he expired on Tuesday morning at half past ten o'clock.

I have the honour to be, Gentlemen,
Your most obediant servant
J.W.K. Parkinson

A motion was carried unanimously that a letter of acknowledgment be sent by the Clerk to the Board, expressing their regret on the occasion of the death of James Parkinson, and "of the high estimation in which his services to the Parish as well as his general character are held by the Board. . . ."

The obituary notices were brief. *The Gentleman's Magazine* merely stated: "In Kingsland-road, James Parkinson, esq. surgeon, late of Hoxton-square" in the list of deaths for December 21, 1824 (Parkinson, 1824).

No. 1 Hoxton Square was still standing in 1910, when it was visited by Professor L.G. Rowntree. He described it as follows:

The house is a plain old three-storey building, facing the east, on the southwest corner of Hoxton Square. Behind the main building and connected with it is a smaller two-storey one with a central door opening into a little side street. This apparently was Parkinson's office. Behind this again is another smaller building which may have served as a laboratory, or a library, or perhaps as a museum. Leading up to the deeply set, black, massive looking front door are a stone walk and deeply worn stone steps. The house is only a few feet back from the street and before it stands an old fence.

Uninteresting though the exterior is, upon entering this building one is impressed at the large size of the rooms and with the evidence of the prosperity of other days. We see in almost every room great carved open fireplaces of elaborate design, and between some rooms large connecting arches. The deep panelling of walls and ceiling which was formerly so much in vogue is well preserved in some of the rooms on the second floor. One is surprised to find such an interesting interior behind such an uninviting exterior (Rowntree, 1912).

A photograph of the frontage was included in Rowntree's article. A factory now occupies the site, but a blue plaque commemorating Parkinson was installed in September 1961. In his will, Parkinson referred to the contents of his shop, so it is probable that he dispensed from Hoxton Square as part of his duties as surgeon-apothecary. The house was never owned by the Parkinson family, the owner being one Joshua Jenning.

Nothing is known as to how or where Parkinson obtained his early education, but it can be inferred from his writings, particularly *The Hospital Pupil* (Parkinson, 1800) that he had received—as he there termed it—a "sound liberal education." He certainly knew Latin and Greek and had a knowledge of French. He became expert in shorthand, with an ability to write at the speed of ordinary speech. Drawing was a favorite subject which he urged medical students to cultivate, and he himself was an accomplished artist who illustrated many of the fossils of his own collection and advised Gideon Mantell on geological illustration. Later, he made a private study of chemistry, stimulated by reading Richard Watson's *Chemical Essays* (Watson, 1781–87), and probably augmented by attendance at occasional lectures on the subject as well as his personal contacts with fellow members of the London Chemical and Philosophical Society.

A little more is known, however, concerning his medical education. He was apprenticed surgeon to his father, and he was one of the earliest medical students at the London Hospital Medical College which he entered in 1776 at the age of 20 and where he studied for a period of six months (Clark-Kennedy, 1962–63).

Some information concerning Parkinson during the formative years of his apprenticeship is to be found in the annual reports of the Royal Humane Society. Founded in 1774, the Society remunerated those attempting resuscitation in cases of drowning, and appointed medical assistants with responsibility for sections of the London Waterways; John Parkinson served the New River area. In 1777, James Parkinson was awarded the Honorary Silver Medal of the Society for the recovery of the body of a Hoxton man who had hanged himself. The report of the case, written by the senior Parkinson, describes the patient's recovery using the restorative methods recommended by the Society and reports further cases of resuscitation undertaken by Parkinson.

James Parkinson has had attributed to him the authorship of *Observations of Dr. Hugh Smith's Philosophy of Physic* (McMenemey, 1955), a work of medical controversy published anonymously in 1780. The author challenged the theories of Dr. Smith regarding the role of Vital Air in the circulation of the blood, as propounded in his popular lectures given in 1778 and published serially in 1780 under the title *Philosophical Enquiries into the Laws of Animal Life* (Smith, 1780).

In the winter of 1784–85, James Parkinson attended a course of surgical lectures delivered by the illustrious John Hunter at his house in Leicester Square. Here, his knowledge of shorthand was invaluable, for he was thus able to take down a verbatim account of the lectures. A transcription of these notes was lent to William Clift F.R.S. (1775–1849), who had them copied in two volumes which are preserved in the Library of the Royal College of Surgeons. Parkinson's notes are invaluable in view of the destruction of a significant proportion of Hunter's manuscripts by Sir Everard Home (1763–1832). Dedicated to Clift, the lectures were published in 1833 by John William Keys Parkinson at his own expense, under the title of *Hunterian Reminiscences* (Parkinson, 1833).

Three months after the death of his father, James Parkinson was examined and awarded the diploma of the Company of Surgeons on April 1,

1784. In 1787 he was elected a Fellow of the Medical Society of London (founded in 1773), being admitted on May 7 at a meeting at which, amongst other notables, Dr. J. C. Lettsom, the founder of the Society, was present. On February 4 of that year, Parkinson had communicated his first paper to the Medical Society, entitled "Some Account of the Effects of Lightning" (Parkinson, 1789), in which he gave a vivid description of the injuries received by two men when a house in Shoreditch was struck.

In the year 1814, Parkinson addressed a letter to the Editor of the *London Medical Repository* (Parkinson, 1814), published under the heading "Cases of Hydrophobia." The letter concerned two cases of hydrophobia which Parkinson had observed some 26 years earlier in conjunction with Sir William Blizard (1743–1835), who had performed the anatomical dissection. Parkinson said that he had given his notes on the case to Dr. Andrew Marshall, a pathological assistant at the London Hospital. These notes appeared unabridged in Dr. Marshall's posthumously published *The Morbid Anatomy of the Brain* (Marshall, 1815).

James Parkinson was one of a number of London doctors, including Henry Cline (1750–1827) and the young Astley Cooper (1768–1841), who openly avowed their support of the cause of parliamentary reform and extension of the suffrage. He was a member of the Society for Constitutional Information, which had been founded in 1780 in order to "restore Freedom and Independence to . . . the legislature" by propagandistic means, and disseminate the principles of reform that underlay the parliamentary bills of the younger Pitt (1782) and the Duke of Richmond (1788).

A nucleus of influential dissenters were early members of the Society, agitating against Pitt's retention of the Acts that precluded dissenters from holding civic offices, and greeting with enthusiasm the events of 1789 in France. By the time of the General Election in June 1790, the political climate had polarised. Members of the Society for Constitutional Information joined with those of the true Whig Revolution Society to commemorate the first anniversary of the fall of the Bastille in the "Crown and Anchor Tavern" in the Strand, whilst in November of that year Burke published his *Reflections on the Revolution in France* (Burke, 1790), which unleashed numerous replies the following year, when the first part of Paine's *Rights of Man* (Paine, 1791) was published. Energetically promoted by the Society for Constitutional Information, the *Rights of Man* provided an important stimulus to the radicalism of literate artisans, who formed the bulk of the membership of the London Corresponding Society, founded (by Thomas Hardy and Maurice Margarot) during the first half of 1792.

Parkinson was an early member of the Corresponding Society, but his contacts with its leaders predated its foundation; from his evidence at the State Trial of John Thelwall, held in the first weeks of December 1794, it is known that his friendship with Thelwall was then of seven years' standing. Parkinson's close associate, the radical publisher Daniel Isaac Eaton, had premises in Hoxton between 1786 and 1792. Parkinson wrote some dozen political tracts which appeared during the years 1793–95, the opening stages of the war with France and the time of increasing governmental attempts to infiltrate and suppress the activities of the radical societies. These pamphlets appeared chiefly under the pseudonym "Old Hubert" (identified as Parkinson in an advertisement in the *Morning Chronicle* of 9/12/1794) and were published either by Eaton or by John Smith, booksellers who were repeatedly, though not always effectively, prosecuted for publishing seditious libel. In his examination before the Privy Council, James Parkinson admitted to Pitt and Eldon that he was the author of two anonymous tracts published under the auspices of the Corresponding Society during the period just before the Treason Trials of 1794: *Revolution without Bloodshed—or Reformation preferable to Revolt* (Parkinson, 1794a), and *A Vindication of the London Corresponding Society* (Parkinson, 1796). In addition, Parkinson contributed *Old Hubert's History of England* (Parkinson, 1794a) to Eaton's magazine *Politics for the People* (Eaton, 1794), as well as a number of short squibs signed "Old Hubert" and extracts from approved political theorists. An anthology of such extracts, *Pearls cast before Swine* (Parkinson, 1793a), appeared in 1793 together with *An Address to the Hon. Edmund Burke from the Swinish Multitude* (Parkinson, 1793): the two parts of *Budget of the People* (Parkinson, 1793b); *The Knave's Acre Association* (Parkinson, 1793a); and the popular *Village Association* (Parkinson, 1793c), which was anthologised and reprinted in extract. *Mast and Acorns* (Parkinson, 1794b); *Whilst the honest poor are wanting bread* (Parkinson, 1795); and *An account of some Peculiar Manners and Customs of the People of Bull-land, or the Island of Contradictions* (Parkinson, 1795c) date from 1794 or 95.

Following the arrests of Thelwall and many others during May and June 1794, Parkinson, together with John Smith and George Higgins, became members of the committee of correspondence of the London Corresponding Society. With Dr. Robert Crossfield, Smith and Higgins were implicated in the Pop-Gun Plot to assassinate George III, in connection with which Parkinson appeared before the Privy Council and gave evidence as a witness for the defence at the trial of Crossfield for High Treason, which began on May 11, 1795. A verdict of "not guilty" was returned, and Parkinson published his account of the events on August 29, 1795 (*Assassination of the King—a Letter to Mr. John Smith*). By this time, Parkinson had acquired a considerable degree of public notoriety: "REPUBLICAN, DEMOCRAT, and LEVELLER, are distinctions with which I have for some time been favored, in common with every one who had reprobated the present accursed war, waged against mankind. . . . As I fully believe that—*Titles without merit are truly despicable*, I experienced a considerable degree of exultation on reflecting that *titles*, so *absolutely honourable*, were not bestowed on me without my having merited them."

With the acquittals of the defendants in the Treason Trials of 1794 and 95, the membership of the Corresponding Society increased dramatically, the radical movement reaching its peak of activity. On November 6, 1795, Grenville introduced the Treasonable Practices Bill; on the 10th, Pitt moved the Seditious Meetings Bill and, despite intense protest, the Two Acts became law on December 18. Their effect was to severely curtail the public activity of the societies, which were forced underground and into clandestine association with the more extreme republicanism of the United Irishmen. Some of Parkinson's associates remained active at this late stage; Smith was sentenced on February 11, 1797 to two years' hard labor for libellously publishing *A Summary of the Duties of Citizenship.*

In 1807, Samuel Whitbread, M.P., son of the brewer and member of the Board of Trustees of the parish of St. Luke's, introduced a bill in the House of Commons "for providing and encouraging industry amongst the labouring classes of the country and the relief and regulation of the criminal and necessitous poor." Amongst other measures, it sought to establish parochial schools throughout the country, providing free education for pauper children for a period of two years. Parkinson was one of a number of authors, including the Rev. T. R. Malthus (1766–1834), who criticised the bill. Parkinson had been appointed secretary to the Sunday School held at St. Leonard's Church, and an account of the work and aims of the school is given in his tract "*Remarks on Mr. Whitbread's Plan for the Education of the Poor*" (Parkinson, 1807).

Whilst a debt of gratitude was due to the proposer of the bill, Parkinson feared that, as a result of the establishment of the parochial day schools, attendance at the Sunday Schools would suffer and "that a mere literary education without the inculcation of moral and religious principles would prove highly injurious." As Sunday School superintendent, he introduced a register containing the names of all children seeking employment, with particulars concerning their abilities and behavior, which was open to inspection by prospective employers. One pragmatic criticism of Whitbread's provisions was that the parents of children in employment would be unlikely to forego their income for the sake of their education. But Parkinson here also criticised Malthus, stating "that if the population exceeded the means of support, the fault lay not in nature, but in the ability of politicians to discover some latent defect in the laws respecting the division and appropriation of property" (Parkinson, 1807).

In 1799 and 1800 Parkinson published five works, but some of these may have been awaiting publication during the imprisonment of John Smith. Two of the books were on the subject of domestic medicine, namely, *Medical Admonitions to Families* (Parkinson, 1801) and *The Villager's Friend and Physician* (Parkinson, 2nd ed., 1804), which ran into several editions. *Dangerous Sports* (Parkinson, 1808) consisted of a sequence of stories for children, warning against the playing of dangerous games; these were somewhat in the style of Thomas Day's *Sandford and Merton* (Day, 1783–89) or the children's books published earlier by John Newbery. *The Hospital Pupil* (Parkinson, 1800) printed letters from Parkinson to a father whose son desired to become a doctor, and outlined those qualities he considered requisite to the medical student, providing a sketch of a suitable course of study. The book was well received by the medical press, and a second edition was issued in 1817. The first of what were to be four editions of *The Chemical Pocket-Book* (Parkinson, 1800a) was published in 1800. Continuing his series of popular medical works, in 1802 Parkinson issued a broadsheet entitled: *The Way to Health* (Parkinson, 1802), designed to be framed and hung by the cottager's fireside as a reminder of the laws necessary to maintain health and sobriety. *Hints on the Improvement of Trusses* (Parkinson, 1802a) of the same year was cheaply priced to reach the poorest afflicted with inguinal hernia, and described how to construct at small expense a truss of his own invention. Parkinson suffered from gout, and had been afflicted with what he called "that tormenting malady" for 15 years before he wrote *Observations on the Nature and Cure of Gout* (Parkinson, 1805). It is notable both for Parkinson's detailed account of his personal experience of the disease and for his re-adoption of the role of medical controversialist, as the work attacks the refrigeration treatment of gout advocated by Dr. Robert Kinglake of Taunton. It elicited a spirited rejoinder in 1807, when Kinglake published his *Strictures on Mr. Parkinson's Observations* (Kinglake, 1807).

During the latter part of his life, "Mr. Parkinson, an English naturalist" attained celebrity for his researches on fossils. He said that he "always had an insatiable curiosity to pry into the mysteries of the natural world," and as a student of oryctology, the former term for palaeontology, he amassed one of the largest and most valuable collections of fossils in Britain. His specimens of Organic Remains were displayed in cabinets at his house in Hoxton Square and attracted numerous visitors, amongst whom was Dr. Gideon Algernon Mantell, parish doctor and army surgeon of Lewes, Sussex, and later a celebrated geologist. Mantell has acknowledged that he owed much to Parkinson's instruction and encouragement in his formative years.

In the absence of a portrait of Parkinson, it is to Dr. Mantell that we owe the only description of his appearance and personality. "Mr. Parkinson," wrote Mantell, "was rather below middle stature, with an energetic intellect, and pleasing expression of countenance, and of mild and courteous manners; readily imparting information, either on his favourite science, or on professional subjects."

Between the years 1804 and 1811, James Parkinson published the three quarto volumes that comprise his *Organic Remains of a Former World* (Parkinson, 1804–11). The completed work was a comprehensive study of palaeobotany as well as of palaeozoology, bringing together the piecemeal discoveries of the 18th century, such as those of John and William Hunter, yet retaining its Biblical explanatory framework, as the title makes explicit. But, as with other contemporary work, Parkinson emphasised the importance of oryctology to geology, rather than as an appendix to botany and zoology, and tried to link palaeontology to the study of strata. As a founding member of the Geological Society of London, established on November 13, 1807, he contributed to its *Transactions* one of the earliest accounts of the strata of the London district and its fossils (Parkinson, 1811a). The *Transactions* contain two further papers by Parkinson, dating from 1814 and 1818 (Parkinson, 1814a and 1818); he also contributed two geological papers to *Nicholson's Journal* in 1809.

In the year 1822 he published his geological manual, *Outlines of Oryctology, an Introduction to the Study of Fossil Organic Remains* (Parkinson, 1822). In recognition of his contributions to geology, Parkinson was elected an honorary member of the Wernerian Society of Natural History of Edinburgh, and the Imperial Society of Naturalists of Moscow.

For more than 25 years Parkinson was the regular medical attendant at Holly House, Hoxton, a private madhouse of 118 beds which admitted both fee-paying and pauper patients. On their visits the Commissioners in Lunacy reported overcrowding, particularly in the pauper divisions; yet, of the three well-known Hoxton madhouses, it seems to have provided the most humane treatment for its inmates. In 1810, James Parkinson was a medical witness at the trial of a man who was found guilty of having conspired to confine his aunt, Mary Daintree, when it was alleged she was of sound mind. Parkinson was severely reprimanded for having certified the woman as insane, not as a result of his own observations but merely on the report of her relatives. Parkinson's conduct was attacked in the *Statesman's* accounts of the trial, and also questioned in the medical press. This censure must have harmed his professional reputation and personal integrity. Parkinson took pains to vindicate his conduct in *Observations on the Act for regulating Madhouses* (Parkinson, 1811b), which he published in 1811.

When the Royal Jennerian Society, formed in honour of Dr. Edward Jenner for the purpose of encouraging vaccination, established some 13 vaccination stations in London, Parkinson was appointed Inoculator at the station on the premises of the Hoxton Academy Chapel Sunday School in Hoxton Street. Parkinson was a member of the medical council of this society, and with some 20 other doctors he was a signatory to a report of the effectiveness of vaccination, submitted to the Society in 1806.

At the beginning of the 19th century, Messrs. Parkinson & Son must have had a large and lucrative practice, for the population of Shoreditch had increased from 30,000 inhabitants in 1801 to 50,000 in 1811. This number were served by but half a dozen qualified apothecaries and surgeon-apothecaries, although quacks abounded. The influx was chiefly working-class. After James Parkinson's death, J. W. K. Parkinson was to

Figure 4. Holly House, commonly known as Burrow's Madhouse in Hare Street, Hoxton.

complain that parish patients were occupying far too much of his time, and that in having to make some 70 to 80 domiciliary visits per day to sick paupers, he was finding difficulty in attending private midwifery cases, which he said were the most profitable part of his practice.

John Parkinson was the author of two papers which were read before the Medico-Chirurgical Society of London by his father, and afterward published in the Society's *Transactions*: "A Case of Trismus successfully treated," and "Case of a diseased vermiform appendix." The latter paper provided an early description of appendicitis from a post mortem examination (Parkinson and Parkinson, 1811 and 1812).

In 1813, "Parkinson & Son" were appointed "Surgeon, Apothecary, and Man-midwife" to the poor of the parish of St. Leonard, at a salary of £156 per annum. As parish doctors, they were responsible for the medical care of paupers in the workhouse and the workhouse infirmary, as well as outdoor paupers on medical relief whom they attended either in their own homes or at the parish dispensary. In addition, they had the care of pauper infants at the workhouse nursery, which was situated at Enfield. Some 20 to 30 pauper lunatics who were boarded out by the parish at Sir Jonathan Miles' private madhouse had also to be visited regularly. After a year as parish doctor, Parkinson complained that his salary was inadequate for his services; the Board of Trustees promptly raised his salary to 250 guineas per annum.

His first step as surgeon to the infirmary was the classification and segregation of inmates in adult, children's, maternity, imbecile, and observation wards. In 1814, a severe epidemic of the putrid fever (typhus) occurred in Hoxton, with over 100 cases in the Kingsland Road alone. Due to overcrowding of the infirmary, he persuaded the trustees to permit him to send further cases to the London Fever Hospital at the expense of the parish. The Board also accepted his recommendation that a separate fever block should be added to the infirmary for the segregation of fever cases, but five years were to elapse before the fever wards were built. A tract entitled *Observations on the necessity for Parochial Fever Wards* (Parkinson, 1819), published in 1819, contains his arguments for the isolation of fever cases. In a subsequent epidemic of typhus which occurred in the parish after the fever wards had been built, the mortality was substantially lower, according to the joint paper of James and J. W. K. Parkinson: *On the treatment of Infectious or Typhoid Fever* (Parkinson, 1824).

Parkinson was a member of the Medical Society of London (from 1787); the Medico-Chirurgical Society of London (his name appeared in the first list of members published in 1812); and the Hunterian Society, of which he was a founding member (1819). He was an early member of the Society for the Relief of Widows and Orphans of Medical Men, founded in 1788. He joined the Association of Apothecaries and Surgeon-Apothecaries of England and Wales. As a member of the council of the Association, he was one of the instigators, through the Society of Apothecaries, of the Apothecaries Act of 1815. This Act provided a legal foundation for their professional status by enabling quacks to be prosecuted and by instituting examinations for the diploma of Licentiate of the Society of Apothecaries. In 1817 James Parkinson was elected President of the Association, an office which he held for three years.

An Essay on the Shaking Palsy (Parkinson, 1817) was published in 1817, when James Parkinson had reached the age of 62 and had behind him a wealth of clinical experience, gathered during more than 32 years of private and institutional experience. His clinical acumen and power of observation, coupled with his characteristic descriptive skills, resulted in the writing of what has justly become a medical classic. Parkinson had been attracted to a study of what he termed "a tedious and most distressing malady" (Parkinson, 1817, p. 12) by a case which he had under observation for a number of years, the first described in his series of six cases. All six cases had involuntary tremors of the limbs, varying only in degree and distribution. Five of them had the further characteristic of a "propensity to bend the trunk forwards, and to pass from a walking to a running pace" (Parkinson, 1817, p. 24). Although both of these symptoms had previously been observed, they had hitherto been regarded as distinct nosological species. Parkinson's originality lay in his recognition that these two components were pathognomic of a single malady, which he termed "Shaking Palsy, Paralysis Agitans." The Essay was well received at the time, being noticed by several literary periodicals as well as by medical reviewers. Parkinson's classical clinical description of Paralysis Agitans became gradually accepted during the 19th century, both in Britain and on the continent. In the 1860s in his accounts of his investigations into the disease at the Salpètriere with Vulpian, Charcot canonised Parkinson's achievement in the eponymous term "la Maladie de Parkinson." Its anglicised form, ironically, was but tardily accepted.

CHAPTER 3

The Political Radical

Since the sole object of the London Corresponding Society was to obtain reform of the parliamentary representation of the people, which had in turn helped lay the foundations of democracy in Great Britain, James Parkinson's participation is of some importance. The leaders of the Society were intelligent, honorable, and worthy men. Though execrated by the political party in power and indeed by the nation at large, their characters have been cleared of misrepresentations and calumnies, and they have merited the gratitude of a posterity now accustomed to democratic government.

To understand this background, the social, political, and economic conditions of the period require evaluation. King George III had occupied the throne for more than 30 years, the country was prosperous and wealthy, there was relatively little unemployment, the cost of living was low, and wages reasonable. In 1792, when the Society was formed, taxation had been increased so much that it became a burden greater on the poor than the rich. Wars had drained the country's resources to a perilous degree. Discontent and disaffection with royalty, aristocracy, Ministers of the Crown, and the hangers-on and sycophants known as placemen and pensioners, were apparent, a state of affairs consequent on the declining standard of living, and the increasingly apparent inequalities of the laws as applied to rich and poor. Equalisation of the laws was necessary and the first step would need to be equal representation in the House of Commons.

Parliament had long been aware that it was not truly representative, and as early as 1766, William Pitt the Elder, later Earl of Chatham (1708–1778), had prophesied, "Before the end of this century, either this parliament will reform itself from within, or be reformed with a vengeance from without." William Pitt the Younger (1759–1806) had in 1782 introduced a bill in the lower House to enquire into the electoral representation of the country, which was also rejected. The young Pitt had said, "This House is not the representative of the people of Great Britain; it is the representative of nominal boroughs, of ruined and exterminated towns, of noble families, of wealthy individuals and foreign potentates." The bill, defeated by 161 votes to 141, was reintroduced by Pitt in 1785 but again met a similar fate, and he spoke contemptuously of the borough representations as "the rotten part of our constitution." In 1788 the Duke of Richmond (1735–1806) introduced a bill into the House of Lords for the establishment of annual parliaments, universal suffrage, and equal electoral districts, which was also rejected.

It was about this time that a number of reform societies were founded, chief among them being the Society for Constitutional Information (1780) and the Constitutional Club (1788), but both drew their membership from the upper echelons of society, taking pride in their exclusivity (Goodwin, 1979).

The younger Pitt had founded his bills upon plans laid down by his father, but even its sponsors, the Whigs, were not prepared to sacrifice property and influence. The bill of 1785 would have meant the gradual extinction of all the decayed boroughs and disfranchisement of them immediately. The borough-mongers, as the holders of rotten boroughs were called, were to be compensated at market value. As an indication of the value of "rotten boroughs," Gatton, in Surrey, was sold in 1775 for £75,000. It had three electors—Sir Mark Wood, his son and his butler—and two members of Parliament, Sir Mark and his son. Parkinson wrote of this system: "The seats in that House are bought and sold like Standings for cattle in a fair," and added that Parliament "is disgraced by a band of men sent there by an established system of private patronage, 306 out of 513 of its members being appointed by the Treasury and 162 by individual peers and rich borough-mongers. Traitors! Traitors! Traitors!"

The Revolution of 1688 had given the people a Bill of Rights which ensured freedom of conscience for everyone and the right of the people to govern themselves through their representatives in Parliament. This was good in theory, but in fact the people were not all represented in Parliament. A place like Gatton, Surrey, with three electors, could send two members to the House of Commons and yet the large, thickly populated manufacturing towns had no representatives.

Pressure for reform was mounting—partly in response to a period of peace in which no military distractions served to divert parliamentary attention from uncomfortable home truths. But this was not to last. In France, stirring events were happening and it seemed that a new constitutional era was dawning, not only for France, but for the whole of Europe. When the mob destroyed the Bastille in 1789, the last remnants of feudalism and despotism were swept away.

Louis XVI of France, hoping to save himself, had summoned the States General (which had not met since the days of Cardinal Richelieu) and renamed it the National Assembly, which abolished the privileges of the nobles, their immunity from taxation, and forced the King to accept a new constitution to replace his despotic power.

In England, news of the French Revolution was at first received with genuine delight, and all classes welcomed the change to a new order of constitutional freedom which provided for the liberty of the King's subjects. "How much the greatest event it is that ever happened in the world, and how much the best," was how it was hailed by Charles James Fox (1749–1806), but Pitt was not so enthusiastic, although he thought that law and order in France would adjust in time, and that "France would stand forth as one of the most brilliant powers in Europe."

By the following year, the great majority of Englishmen viewed these changes in a different light. Mistrust and apprehension were soon turned into violent hatred, fanned into flame by Edmund Burke (1729–1797) who, with his impassioned oratory, had swayed Parliament. Burke was, from the very beginning, an uncompromising opponent of the French Revolution, and stood alone in Parliament in his denunciation of the revolutionaries and their methods. Parliament looked upon him as something of a bore—he was growing old, his eloquence was failing, and his esteem was low. So he took to his pen and pleaded that the armies of Great Britain and other European countries should intervene on the continent; Pitt wisely and cautiously opposed any interference of this sort.

In October 1790, Burke published his book *Reflections on the Revolution in France* (Burke, 1790), in which he denounced not only acts of violence, but the ideological underpinning of such radical change. Nonintervention, according to Burke, meant "that the age of chivalry is gone . . . and the glory of Europe is extinguished for ever." In his book, he railed at Pitt and taunted him for his timidity, but Pitt was indifferent to these insults, calmly asserting that he would preserve neutrality unless it became necessary to take up arms in self-defence. Nor did Burke's own party, the Whigs, side with him.

Burke was much encouraged by the success of his book, which sold 30,000 copies within a year. Where his speeches had failed, his pen had succeeded. A deluge of replies poured forth in the form of letters, pamphlets, cartoons, and books, including Thomas Paine's (1737–1809) *The Rights of Man* (Paine, 1791). Burke had scornfully referred to the common people as "the swinish multitude," a remark which attracted the attention of James Parkinson, who wrote a satirical pamphlet called *An Address to the Hon. E. Burke from the Swinish Multitude.* "As for yourself, good Sir," wrote Parkinson, "there are those who contend that you resemble the venomous asp, whose poison is placed beneath its tongue; others compare you to the Camelon, and say that, like that creature, your appearance ever changes with your situation" (Parkinson, 1793).

Burke was gratified by the effect of his book on the populace in stimulating hatred of the revolutionaries, but disappointed with its failure to influence Pitt and the Government. He appealed instead to some refugee princes, who had raised an army to march against France and were now quartered on the Rhone. Burke, who condemned terror on the part of the French, sent his son to join the army on the Rhone, and conveyed to him a message: "Be alarmists. Diffuse terror."

France, in April 1792, declared war on Austria, which had refused France's demand to break up the refugee army. Austria was soon joined by Prussia, and the combined armies proceeded to march against France. War sparked the Reign of Terror in France itself, as if in response to Burke's call for terror against her.

The mob broke into the Tuileries on August 10, 1792, and King Louis XVI was imprisoned. The Paris Commune, instigated by the Jacobin party, whose aim was a republic, had triumphed. A National Convention proclaimed that royalty and aristocracy were abolished and the populace demanded that the nobles be sent to the guillotine. By September 1792, the massacre of the aristocracy had commenced, the people of England were horror-stricken, and the feelings of sympathisers were soon turned to revulsion.

The London Corresponding Society was founded in 1792 by Thomas Hardy (1752–1832), a shoemaker, of No. 9 Picadilly, who had for some time considered the formation of an association of men who had no vote in parliamentary elections. On January 27, Hardy and nine others met at the Bell Tavern in Exeter Street, Strand, where it was agreed to form a society whose aim and object would be to obtain a reform of parliamentary representation with the institution of universal suffrage. Several names were suggested, such as the Patriotic Society, the Reforming Society, etc. Finally, it was decided the name would be the London Corresponding Society, as it was thought desirable to correspond with other societies, in London and the provinces, having similar opinions, in order to ascertain their views on parliamentary representation. Hardy was elected secretary and treasurer, and Maurice Margarot, a merchant of High Street, Marylebone, a "gentleman of elegant manners, a scholar and marked for a leader," was chosen as chairman. Rules and regulations for the conduct of the Society were formulated, the first of which required a candidate for membership to be vouched for by two other members, and a declaration affirmed:

> That he is convinced that the Parliamentary Representation of this country is at present inadequate and imperfect—That the welfare of these kingdoms requires that every person of adult years, in the possession of his reason, and not incapacitated by Crimes, should have a vote for a Member of Parliament—And that he will endeavour, by all justifiable means, to promote such reformation on Parliament.

Rule 3 stated that the Society was to be organised into divisions, consisting of 20 members, and that each division was permitted to enroll additional members to the number of 40 before dividing again to form another division of 20 members. Each division was to appoint one delegate who would become a member of the Committee of Delegates. A division was to meet weekly under a chairman appointed by the divisional members. Rule 5 stated that each member should pay to the Chairman of the division the sum of one penny per week.

A manifesto dated April 2, 1792, explaining the purpose of the Society, was printed and widely circulated. The manifesto made an immediate appeal to the middle and lower classes, especially the latter. Small tradesmen, journeymen craftsmen, shop assistants, industrial workers, and others of the laboring classes rushed to join what was virtually a working class movement. Many professional men, doctors and lawyers, and "gentlemen" also joined. Hardy himself was a protestant dissenter, an upright, honest man, whose character was irreproachable, and he attracted large numbers of dissenters and dissenting clergymen into the newly formed organisation.

Hardy was a strict disciplinarian. Most of the divisions held meetings in public houses, but no smoking or drinking was permitted during the meetings. No matters tending to cause "disunion or disloyalty to the crown, were permitted, and no religious disputes."

"Man," the manifesto proclaimed, "as an individual is entitled to Liberty—it is his birthright. As a member of society the Preservation of that Liberty becomes an indispensable duty."

The manifesto, entitled *An address from the London Corresponding Society to the Nation at large in respect of Parliamentary Reform* declared "that fraud, or force, withheld the right of sharing the Government of the country from the majority of the inhabitants of this country" (London Corresponding Society, 1792). Members of Parliament seldom procured their seats from the unbought suffrages of the people. The consequences of unequal and inadequate representation, together with widespread corruption, were oppressive taxes, unjust laws, restriction of liberty, and waste of public funds. Finally the Society condemned violence; its aim was reform not anarchy, and "reason, firmness and unanimity" its only weapons.

The publication of the second part of Paine's *The Rights of Man* (Paine, 1792), coincided with the formation of the Society. In this volume Paine criticised the hereditary right of the King to govern, praised the constitution set up by the American republic, and condemned the wasteful expenditure of the British government. The book was cheap and therefore within reach of the vast majority of the people, who read it avidly. Paine gave the proceeds, which amounted to £1000, to the London Corresponding Society. The Government thought the book was safe in the hands of the educated classes but quickly saw the danger of its ideology contaminating the lower orders, where it would sow the seeds of discontent and dissaffection for the King and his Government. It was considered a libel on the Constitution and the publishers and sellers were later prosecuted.

The Society was formed as events in France reached a climax. Although its members sympathised with the revolutionaries, popular feeling in the country had turned sharply away.

In September 1792, the Society sent a letter of support to the National Assembly of France. In spite of its loud public utterances of loyalty to King and Constitution, this action drew government suspicion. Matters were made worse when the Society for Constitutional Information sent two members to France with greetings. The Societies were now seen as hornet nests of sedition and revolution.

A few months later, on December 13, 1792, the Speech from the Throne spoke of "an attempt to destroy the Government," and of "measures subversive of all order and government," clearly referring to the activities of the reforming societies. A Royal Proclamation followed, warning the people against seditious publications, and ordering magistrates to prosecute the authors, printers, and sellers of such writings.

A series of prosecutions followed, beginning with the author of the "Rights of Man," but Paine had gone to France, accompanied by Dr. Joseph Priestley (1733–1804), and remained there safe from English jurisdiction. James Ridgway and Henry Delahoy Symonds (both of whom published books written by James Parkinson) were each sentenced to four years imprisonment for selling Paine's book. The Rev. William Winterbottom (1763–1829), a dissenting minister of Plymouth, was sentenced to four years imprisonment for saying in a sermon, "if the King does not observe certain laws and rules he has no more right to the Crown than the Stuarts had." Likewise John Frost (1750–1842), an attorney, a prominent member of the London Corresponding Society and the Society for Constitutional Information, for announcing that "equality is every man's birthright. I am a republican," was sentenced to six months in prison, ordered to stand in the pillory, and was crossed off the roll of attorneys. Daniel Isaac Eaton (d.1814), publisher and printer of a periodical, *Politics for the People* (Eaton, 1794), to which Parkinson had contributed a serial "Old Hubert's History of England" (Parkinson, 1794a), was prosecuted on account of the following extract from an article, "I had a fine majestic sort of animal, a game cock, a haughty sanguinary tyrant, nursed in blood and slaughter from his infancy; fond of foreign wars, and domestic rebellions, into which he would sometimes drive his subjects by oppressive obstinacy, in hope that he might increase his power and glory by their suppression." Eaton was acquitted. *Politics for the People* appeared with the added title of "Hog's Wash," or "a Salamagundy for Swine," but only a few more numbers were published.

Within a year of its inception the London Corresponding Society had achieved a large membership, but after the Royal Proclamation a large number deserted. Joseph Gerrald (1763–1796), a barrister and prominent leader, succeeded in obtaining the signatures of some 7000 members to a petition for submission to Parliament, pleading for reform. But many more members would not append their names in fear of the consequences of publicity, some fearing for their jobs. Tradesmen worried over loss of business, and professional men were afraid their practices would suffer. Charles James Fox refused to have anything to do with the petition since he was not in favor of universal suffrage, but Philip Francis (1740–1818) M.P. very courageously brought forward a bill for reform. The bill was defeated by 282 votes to 4, plainly showing the temper of a house now united against reform. When in the debate it was asserted that reform would free the representation from corruption, a member said that "the expense of purchasing a seat in the House was not a proof of corruption but merely of the wealth and property of the country." In spite of a diminishing membership and adverse circumstances, the Society continued.

In the year 1793, the French army engaged with the combined armies of Austria and Prussia and invaded Holland. In doing so, they broke a treaty signed by the French government only two years previously. England was now being forced to join in the war but Pitt clung to neutrality even though the country, still shocked by the French massacres of the previous year, called for action. Pitt hoped that he might be called upon to mediate between the warring parties. "No hour of Pitt's life is so great as the hour when he stood alone in England, and refused to bow to the growing cry of the nation for war" (Hope, 1982). It was France who plunged England into the war, thus breaking the peace for which Pitt had fought so desperately.

On January 21, 1793, Louis XVI was sent to the guillotine. In the previous November the French National Convention offered its protection to all nations struggling for freedom and offered the aid of its soldiers to all nations who would strive for freedom. "All Governments are our enemies," said its President, "all peoples are our allies." After the death of Louis XVI the British Government broke off all diplomatic relations with the French republic and its Jacobin leaders. Chauvelin, the French Ambassador who had been fraternising with the reform societies, and who had informed his government that Mr. Pitt was about to recognise the French Republic, was considered *persona non grata* and sent back to his native country.

In February 1793, France issued her Declaration of War against England. Pitt, by character, was unfit for the conduct of a war and knew it, but he drifted along, swayed by the patriotic enthusiasm of his countrymen, feeling that all the good he had done was being undone. Taxation, which he had reduced to its lowest limits, now leaped to an undreamt height, and the cost of living rose rapidly. In Spitalfields, where many Hoxtonians were employed, about 5000 weavers were out of work and many families were starving. The sufferings of the people caused fear and panic, and they vented their wrath upon the societies, partisans of the French Republic, considered to be striving for the establishment of a republic in this country. Formal opposition to the doctrines of the societies was organised by the establishment of the "Association for Preserving Liberty and Property against Republicans and Levellers." A manifesto addressed to the public warned the people against the reform societies "instituted with the express purpose of overthrowing the laws and liberties of their country," and asked the public to be watchful and to assist in bringing to justice those who offended by selling seditious and treasonable publications (A.P.L.P.R.L., 1792). It was pointed out that the sellers were equally as liable as the authors, printers, and publishers, and therefore subject to the same penalties laid down by law.

The Government's secret agents now set to work to spread alarm and despondency, using a story of a fancied insurrection. Some 4000 pikes, battleaxes, knives, etc. had been ordered by the London Corresponding Society, and of course the spies knew where they were manufactured, who had ordered them, and when they were to be delivered. Meetings of the Society were being held behind closed doors, guarded by sentries armed with pikes, whilst within the members were being drilled and taught the use of arms. Pitt believed this concocted story of "thousands of bandits" waiting, fully prepared for the day of revolution. When his niece read to him Thomas Paine's words in vindication of the French Revolution, Pitt replied, "Paine is no fool, he is perhaps right; but if I did what he wants I should have thousands of bandits on my hands and London burnt."

In Scotland there were several societies of a similar nature, but they were relentlessly persecuted. Thomas Muir (1765–1798), a member of the Glasgow Society of the Friends of the Constitution and of the People, was charged with High Treason for causing dissaffection and disloyalty to the King and Government and for seditious writings. He was tried on August 31, 1793, found guilty, and sentenced to 14 years transportation to a penal settlement in Botany Bay, Australia. Thomas Fyshe Palmer (1747–1802), a member of the Society of Friends of Liberty in Dundee, was also found guilty of High Treason on a charge of seditious practices, and sentenced to seven years transportation. Both were savage sentences passed by a partisan and brutal judge [Robert Macqueen (1722–1799), the Lord Justice Clerk, afterwards Lord Braxfield] and a "packed" jury. Vigorous protests were made, in the Lords and Commons, disputing the legality of the sentences, but Pitt and his henchmen were adamant in refusing to mitigate the sentences.

In spite of the fate of Muir and Palmer, the Scottish societies proposed to hold a conference of delegates from all the reform societies in Scotland and invited the English societies to do likewise. This conference was to be called a Convention, a term which savored of its French counterpart; such a term was anathema to the anti-Jacobins of Great Britain. The English societies were chary of sending delegates and much argument followed.

At the annual general meeting of the London Corresponding Society of 1793, more than 700 members attended, and it was decided to hold an open air meeting on October 24th, in a field behind the Hackney Road, Shoreditch, the property of a member named Thomas Briellat, for the purpose of choosing two delegates for the Scottish Convention. The meeting was well advertised and well attended, the audience including a large number of unemployed Spitalfields weavers. The Police Office in Worship Street, Shoreditch, anticipated trouble and on the day of the meeting the Justices of the Peace set up headquarters at the Nags Head Public House, Hackney Road, where they assembled some 300 constables to keep law and order and deal with offenders on the spot.

Richard Hodgson, a hatter, of Broadway, Westminster, was Chairman of the meeting, and he and other speakers addressed the assembly through a window at the back of Briellat's house which overlooked the field. Maurice Margarot, chairman of the Society, and Joseph Gerrald were the chief speakers and were chosen to act as delegates to the "British Convention of delegates of the Friends of the People associate to obtain universal suffrage and annual parliaments," to give the Scottish Convention its full title. It is estimated that well over 1000 persons attended the meeting, and apart from trampling down the boundary railings of the field which proved too small to hold the crowd, the meeting was orderly and quiet. The magistrates were disappointed by the peaceful meeting but arrested Briellat on a charge that 10 months previously he had uttered seditious words. He was committed to the Old Bailey, tried, found guilty, and sentenced to a fine of £100 and imprisonment for one year.

The Convention met at Mason's Lodge, Blackfriars Wynd, Edinburgh, in November 1793. The daily sittings commenced with prayer, "asking for divine assistance in the cause of reform," the doors were kept open, and visitors were encouraged to attend. Some 45 Scottish delegates were present, together with two delegates of the London Corresponding Society, Margarot and Gerrald, and one each from the Society for Constitutional Information and from a Sheffield society. Government spies were present at every session and, after the fourteenth sitting Margarot and Gerrald were arrested as they lay in bed in their lodgings at the Black Bull Tavern. William Skirving, the Scottish holder of the Convention, was also arrested.

The three men were charged with uttering seditious words and released on bail. Margarot and Gerrald returned to London whilst awaiting their trial, and though advised that they would appear before the brutal Lord Justice Clerk and his specially chosen jury, only to meet the same fate as Muir and Palmer, they did not flee the country. Gerrald, a brilliant advocate and eloquent speaker, was convinced he could argue his case successfully and Margarot was equally confident. They returned to Edinburgh where they were tried on March 14, 1794. Gerrald immediately objected to being tried by this particular judge but his words fell on deaf ears. The result was a foregone conclusion; Margarot, Gerrald, and Skirving each received a sentence of 14 years transportation. Protests were again raised in the Lords and Commons asking for a revision of the savage sentences, but to no avail.

Misfortune was now descending upon the London Corresponding Society. Pitt, in common with the nation at large, believed that the country was in imminent danger from within. The press was blamed for sowing sedition and so were dissenting ministers, but the Government thought that the greatest danger lay in the reform societies.

On May 2, 1794, the Society for Constitutional Information dined at the Crown and Anchor Tavern, Strand, with John Wharton, M.P. in the chair. Toasts were drunk to the "Rights of Man," "A Free Constitution," and "The

Swine of England." As a result a bill was brought before the Commons against seditious practices and assemblies, restricting the holding of public meetings and giving wider scope to the Statute of Treasons. Pitt, in a long speech in the House, spoke of a dangerous conspiracy being on foot; the country was on the verge of catastrophe, and he laid the blame squarely on *The Rights of Man*. It was that book which was at the root of it all. He called the attention of the House to a society composed of despicable and contemptible, though for the most part talented and educated, persons. He concluded this speech by sounding the tocsin. He asked leave to bring in the Habeas Corpus Suspension Bill which passed into law on May 22, 1794, having been approved by no less than 250 votes to 39, thus extinguishing the most cherished liberty of the British subject.

On May 12, 1794 the blow fell. Thomas Hardy was arrested and sent to the Tower of London. The arrests of John Horne Tooke (1736–1812), John Thelwall (1764–1834), John Augustus Bonney, Stewart Kyd (d.1811), Jeremiah Joyce (1763–1816), Thomas Wardle, Thomas Holcroft (1745–1809), John Richter, Matthew Moore, Richard Hodgson, John Baxter, and John Lovett followed soon after. All excepting Lovett, who was discharged, and Wardle, Moore, and Hodgson, who were never in custody, were sent to the Tower. All were leading figures and members of governing committees of the two societies (Hope, 1982).

Hardy was tried first. His trial took place at the Old Bailey on October 28, 1794 before the Lord Chief Justice, Sir James Eyre (1734–1799). The Attorney-General, Sir John Scott (1751–1838) prosecuted, and the Hon. Thomas Erskine (1750–1823) defended. The charge was High Treason, to which Hardy pleaded not guilty.

The mass of evidence against Hardy was the testimony of agents-provocateurs who were members of the London Corresponding Society. Since the society for Constitutional Information consisted of but 50 or 60 members, all known to each other, whose integrity was inviolable, the spies could not penetrate it. George Lynam, said to be an ironmonger, was a member of the London Corresponding Society and a paid Government informer. On cross-examination it was proved that he had no real occupation apart from his spying activities and, though he took notes at the meetings, he could show nothing to prove that Hardy had ever expressed disrespect of the King and the Government. The Privy Council in their secret enquiries had been warned about Lynam and distrusted him and his "affectation of knowing some dreadful plot."

Groves, an attorney in some manner attached to Bow Street Police Office, and Gosling, a Bow Street runner, were some of the many informers who stated that pikes, battleaxes, and knives were ordered in large quantities by the London Corresponding Society. It appears that at this time there was a Loyal Lambeth Association founded in 1792, which was a "proper military association to defend our country, lives and property." The members drilled, wore uniforms, and were instructed in drill and the use of arms, but there was no secrecy about it. Unfortunately, as it was later discovered, some members of the London Corresponding Society were also members of the Lambeth Association, thus providing the spies with the evidence they were looking for, since the Lambeth Association was "unlicensed and unauthorised." A few members of a Birmingham Society did, however, arm themselves in self-defence after the attack on the house of Dr. Joseph Priestley. Robert Watt, a member of the London Corresponding Society but another spy, had put before the Lord Advocate of Scotland the story of a proposed insurrection complete with details regarding a supply of weapons. But, as he wanted the sum of £1000 for the information, he was watched by counterspies who discovered that the agent-provacateur had himself "proposed treasonable plans" at the meetings. He was tried in Edinburgh before the notorious Lord Justice Clark, and sentenced to death. In sentencing him the Judge said that he wished the penalty for High Treason—hanging, drawing, and quartering—had not fallen into disuse. Before he was hanged Watt made a full confession of his intrigues.

Amongst the witnesses for the defence at Hardy's trial were Richard Brinsley Sheridan (1751–1816), M.P., Philip Francis, M.P., and Earl Stanhope (1753–1816). James Parkinson was subpoenaed to appear for the defence, but was not called to the witness box.

William Pitt was himself called as a witness for the defence. In cross-examination he would not commit himself to any positive statement that he and Horne Tooke had ever been associated in advocating the reform of Parliament in bygone days. In fact, he could not conveniently remember, nor would he admit, that his own struggle for reform was on a par with the actions of the societies. When Mr. Sheridan gave a frank and clear account of the meetings of the Constitutional Club of 1788, of which he, Pitt, and Horne Tooke were members, Mr. Pitt's memory suddenly revived and he begged leave to add something to his statement, but it was too late as he had left the witness box.

The more light thrown upon the case, the darker appeared the conduct of those who had urged the prosecution, and it became clear "that the personal liberties of the English people were in the gravest peril." The Government men were cocksure that the verdict would be "guilty" and, in anticipation of the apprehension of some 800 members of the societies, it is said that they had prepared warrants of arrest.

The trial lasted eight days. On November 5th the jury retired, and was absent only eight minutes before returning and delivering their verdict of not guilty. In the court was heard "a roar, or rather a convulsive scream of joy," which soon spread to those waiting outside. It was a popular victory. Hardy and Erskine, who had defended him so ably, were transported in carriages drawn by their supporters through the streets of Westminster to their homes. The Attorney-General, on the other hand, himself described how he was hissed as his carriage proceeded along the Old Bailey and down Ludgate Hill.

The Government was perplexed. The result of the case was unexpected, and it had revealed a weak and trivial prosecution to the view of the whole country. Nevertheless it proceeded with the prosecution of John Horne Tooke, which took place at the Old Bailey on November 17th and lasted until November 22nd.

The prosecuting counsel found Horne Tooke a tougher adversary. He was 60 years of age, much older than Hardy, who was 43. He was a doughty debater, self possessed, of intrepid courage, confident, and sufficiently learned in the law to see points in the case as clearly as any of the lawyers. He could argue and present his own case as well as any advocate. Hardy, on the other hand, was a modest man, shy and retiring, not at his best in arguing and pleading his own case. Tooke, it was said in his favor, had always suppressed the extremists in the societies and had prevented them from acting imprudently or impetuously; he had refused election as a delegate to the Scottish Convention. The verdict was not guilty.

On December 1 Bonney, Joyce, Kyd, and Holcroft were put on trial, but no evidence was offered against them, so they were acquitted. Two weeks later Richter and Baxter were similarly acquitted and the others were never brought to trial. John Thelwall was tried at the Old Bailey on December 5 and his trial lasted five days. He also was found not guilty. James Parkinson gave evidence of character at the trial and was examined by Thomas Erskine, counsel for the defence.

The general feeling following the State Trials of 1794 was one of relief. The acquittals of the leaders of the societies proved that the terror was over, and the panic which had seized the country subsided. In January 1795 the Society for Constitutional Information agreed to suspend for the time being all proceedings in furtherance of parliamentary reform. This society never met again. The London Corresponding Society, however, continued to meet and it became stronger than ever, since it was well organised, its aims and objects clearly defined, and its rules strict.

Francis Place (1771–1854), joined the Society on June 8, 1795 and not long after was appointed Secretary. The former leaders did not take up office after the State Trials, though they continued their membership. The moderates, as will be seen later, were gradually replaced by a radical element.

During 1795 the Allied armies had suffered repeated defeats at the hands of the French. Everywhere, except at sea where the Royal Navy was supreme, the French army was successful. The downfall of the Jacobin rulers of France was heralded by the execution of Robespierre, and the Republic was replaced by the Directory, a more moderate Government with whom Pitt was quite prepared to negotiate for peace. Burke, however, again foiled Pitt's attempts in 1796 to end failure abroad and the sufferings of the people at home. He published *Letters on a Regicide Peace* (Burke, 1796), in which he denounced Pitt's efforts in suing for peace and called for a warlike spirit. The years 1797 and 1798 were critical; the continental armies who were British allies had been utterly routed and England stood alone in the struggle with France, with Napoleon extending French conquests on the battlefields of Europe. Conditions at home went from bad to worse. The nation's credit was at a low ebb, a mutiny in the Fleet, a run on the Bank of England, insurrection in Ireland, and threats of invasion by the French, caused alarm and despondency throughout the country. In the midst of such dire events Edmund Burke died. Pitt offered peace to the French Directory but was rebuffed.

In November 1795 two Acts of Parliament were passed which sought to crush the London Corresponding Society. The first provided for the safety and preservation of his Majesty's person and Government, and the second for more effectually preventing seditious meetings and assemblies. It was now virtually impossible to hold meetings except clandestinely. Most of the influential members in the Society resigned and the last meeting was held on December 7, 1795.

In April 1796, Dr. Robert Thomas Crosfield, a medical man and a member of the Society, was charged with High Treason; it was alleged that he had conspired with other members to assassinate the King. He was found not guilty.

Nothing was heard of the Society until the following year, 1797, when Dr. Crosfield became Chairman of the Society and Thomas Evans, a bookseller of Paternoster Row, the Secretary. There followed more trouble, described below, as a result of which Crosfield, Evans, and all members of the committee were arrested.

Eventually an Act of Parliament was passed suppressing all societies "of United Englishmen, Scotsmen, Irishmen and the London Corresponding Society," and at the same time Habeas Corpus was again suspended. The prisoners were never brought to trial but were kept in protective custody for an indefinite period, scattered in provincial prisons, some being detailed for as long as two or three years before eventual release. Apart from not being permitted to have any communication with the outside world, they were well treated. The epilogue was the passing of an Act of Parliament indemnifying Mr. Pitt against all consequences resulting from actions for damages instituted by members of the reform societies.

Each year, on the anniversary of the acquittal of Thomas Hardy, an annual dinner was held to commemorate the event, at which former members of the Society gathered to honor him. At the 28th anniversary dinner held in 1822, 200 persons were present, 24 of whom had been members of the general committee, "all now in business, all flourishing men, some were rich: all giving the best education to their children. The society had been to a considerable extent the means of inducing them to acquire knowledge themselves, and the consequence was their bringing up a race of men and women as superior in all respects to what they would have been as can well be conceived."

Hardy had suffered much. After his arrest a mob attacked his house whilst he was imprisoned in the Tower, and his wife, who was pregnant, was dragged out through a back window in order to escape the violence. She died in childbirth a few weeks later. Hardy's trade as a shoemaker declined and he had to close his shop. Well-wishers set him up in business again in Covent Garden; this also failed. Finally he eked out an existence in a shop in Fleet Street until he was provided with an annuity by Sir Francis Burdett (1770–1844), which kept him for the last nine years of his life. Fortunately he lived to witness the passing of the Reform Act of 1832, but died soon after.

The passer-by, from the pavement, can still see Hardy's tombstone in Bunhill Fields, although the inscription thereon is blurred, eroded by time and weather. On one side are these words: ". . . triumphantly acquitted by an honest and independent jury on the 5th of November, 1794, by which the corrupt and sanguinary ministry of Mr. Pitt was defeated, and a brighter era commenced in the political condition of the country." On the reverse side of this inscription, "Thomas Hardy lived to see a great part of his laudable and enlightened objects fulfilled by the passing of the Reform Bill, which will ultimately lead to a good and happy government. His memory will be cherished by every friend of freedom, piety, and moral rectitude. It will be recorded in the history of this country, that by his excellent conduct through a long life, he demonstrated that the most humble in society, when guided by integrity, and aided by perseverance and judgment, are sure to aid happiness, and advance the liberties of mankind."

Thomas Hardy left a *Memoir* (Hardy, 1832) written in the last year of his life. In the preface he says of the Society, "The London Corresponding Society did more in the eight or nine years of its existence to diffuse political knowledge among the people of Great Britain and Ireland than had ever been done before. . . . Its members devoted themselves to the cause of justice and humanity. They labored zealously, intrepidly, and honestly, although they beheld the guilty arm of power suspended over their heads and ready to crush them, in order to promote the happiness of their fellow citizens" (Hardy, 1832, Preface).

William Pitt, having quarrelled with the King, had resigned from his ministerial office and shortly before he died in 1806, observed, "that he never denied the necessity for some reform in Parliament but objected to the time of commencing it." He had lost the affection of the lower middle classes because he had persisted in repressing freedom of speech and opinion.

Although the panic-stricken reaction against the French Revolution had been forgotten, an agitation for reform rocked the country again in 1809 and Sir Francis Burdett brought in a bill for parliamentary reform which, however, received the support of only 15 members. A pamphlet, in which he criticised the Constitution of the House of Commons, was deemed a seditious publication and Sir Francis was sent to the Tower where he remained until the prorogation of Parliament (Burdett, 1809). Innumerable petitions reached Parliament on this occasion from parish vestries, including one from the parish of St. Leonard's, Shoreditch, which had been drawn up at the request of the inhabitants of Shoreditch by James Parkinson and other members of a specially appointed committee.

At the passing of the Reform Bill of 1832 it was acknowledged that the movement for reform of Parliament owed much to the agitation for reform instituted by the London Corresponding Society, which had instilled into the minds of the common people the necessity for such reform. The Thomas Hardy Anniversary Dinner was held up to the year 1842, by which time only a few of the former members were still alive.

CHAPTER 4

The Pop-Gun Plot

The so-called Pop-Gun Plot involved five members of the London Corresponding Society: John Smith, a bookseller, of Portsmouth Street, Lincoln's Inn Fields; George Higgins, a chemist's shopman, of Fleet Market, nowadays Fleet Street; Paul Thomas Le Maitre, a watchmaker, of Denmark Street, Soho; Thomas Upton, also a watchmaker, of Bell Yard, Temple Bar; and Robert Thomas Crosfield (1768–1802), of no fixed abode. Although all were arrested after an investigation by the Privy Council, Crosfield alone faced the ordeal of a trial for high treason. Parkinson appeared for the defence both before the Privy Council and at the trial.

Crosfield came from a landed family in Yorkshire, studied medicine at Leyden, and graduated M. D. there in 1791. Instead of setting up in practice he devoted himself to drinking and opium taking, and ran up debts on such a scale that he had to surrender his estate to his creditors. He was vain and boastful, adventurous and reckless; when summoned to appear before the Privy Council, he ran away to sea. That no effort was made to arrest him may lend credence to the suggestion that he was spying on the activities of the London Corresponding Society for the government. If this was the case, the motive must have been financial, since his views were always radical and hostile to the status quo.

The story of the Pop-Gun Plot began on the afternoon of September 8, 1794, when Crosfield and Peregrine Palmer, an attorney at the Inner Temple and Crosfield's oldest friend, called at Thomas Upton's shop to collect a watch which had been left for repair. As they left the shop, Upton joined them, saying he was going to Smithfield; Crosfield agreed to walk in the same direction. On the way the three called at the premises of three different brass-founders; in New Street Square, Fetter Lane, in Shoe Lane, Fleet Street, and in Cock Lane, Smithfield. Upton and Crosfield wanted the craftsmen to construct a brass tube, but as they were unable to get what they wanted, moved on. Palmer either stood aside, or waited outside, and was not interested enough to ask what it was they were after. They continued on to Bartholomew Close, and called at the shop of John Hill, a wood-turner, where Palmer heard Upton and Crosfield giving instructions for the construction of a model of a cylinder from which a brass tube could be constructed, as well as an arrow to fit it. Drawings of the apparatus were produced by Upton and explained by Crosfield, but still Palmer was not inquisitive enough to ask to what use the instrument was to be put, until he read in the newspapers of the arrest of Smith, Higgins, and Le Maitre on a charge of having manufactured an air-gun, which was intended to be used to assassinate the King. Palmer then became alarmed and questioned Crosfield, who told him that the model they were after was a wind-gun, but solemnly denied that he knew anything of the purpose for which it was designed.

Upton only a few weeks earlier had offered his services to the Government as an informer. He falsely declared that John Smith and George Higgins had been, with him, actively engaged in searching for someone to construct the weapon, and that Le Maitre had called at his shop to find out if the instrument was ready. Le Maitre's call was witnessed by a female lodger but her evidence was perjured, and a letter from Le Maitre regarding the weapon was forged.

Mr. Robert Ward, an attorney of the Inner Temple, had been deputed by the Government to take a deposition from Upton, which took several days to complete. Ward had challenged Upton's statements, which he began to suspect were a fabrication since there were so many contradictions in them. When Upton asked for a copy which he could retain, Mr. Ward resolutely refused the request.

Upton became apprehensive on hearing this and remarked to Mr. Ward "You consider me then as a culprit." The deposition stated that letters of an incriminating nature had been received from Smith, Higgins, and Le Maitre, but Ward said it was noticeable that when he asked to see these letters their production was delayed for several days. Higgins' letter alluded to his preparedness to manufacture a mixture of lime and mucilage wherewith the barbed arrow could be tipped with poison. Upton had never seen Le Maitre's handwriting, but having challenged Le Maitre to a duel following his expulsion from the London Corresponding Society, a motion which Le Maitre had proposed and Higgins seconded, he received a written reply from Le Maitre which gave him the prototype for the forgery. All the letters shown to the Privy Council were clever forgeries, probably executed by Upton himself.

Interrogated by the Privy Council, Upton had explained the drawings of the air-gun and barbed arrow in such considerable detail as to lead the members to suspect that Smith, Higgins, and Le Maitre had no knowledge as to how the weapons were to be constructed. When challenged he said that he only pretended he knew how the instruments were made and how they would work, that in fact he knew just enough to make the alleged conspirators believe he understood. This plausible explanation, however, overestimated the credulity of the Council, and Upton followed the others to prison.

Crosfield's part in the drama was now revealed. It was he who had drawn the sketches of the weapons which Upton, a skilled mechanic, had devised, and had actively assisted in explaining the requirements to the craftsmen concerned.

What had become of Crosfield? Before leaving London, he had hinted to his friend Peregrine Palmer that he knew the charge against the three men was "false and malicious," and that he was better out of the way as "the charge was unfounded." On February 13, 1795, he had signed on as a surgeon on a South Sea whaler, the *Pomona*, and joined the ship at Portsmouth. England was at that time at war with France and, within a few hours of sailing, the *Pomona* was captured in the English Channel by a French frigate, *La Vengeance*, and taken as a prize to Brest, where Crosfield and the crew were made prisoners of war and transferred to a prison ship in the harbour.

It was reported to the Privy Council on April 27, 1795 that every possible effort had been made to discover and apprehend Crosfield but without success, in spite of the large sum offered as a reward for information which could lead to his arrest. It was believed that he had gone abroad and was now in Lisbon. The Attorney-General gave his opinion that since there was very little probability of capturing Crosfield, it might be in order to release Smith, Higgins, Le Maitre, and Upton. Accordingly, at their next meeting on May 1, 1795, it was ordered that the four men be let out on bail. They were to be charged with high treason and would have to appear for trial on the first day of the Trinity term, before the King's Bench. The three innocent men had been kept in custody from September 28, 1794 until May 8, 1795 without being brought to trial or knowing what their fate would be. Meanwhile the search for Crosfield was to continue.

Crosfield, together with other prisoners of war, was, under an exchange agreement between England and France, repatriated. He now changed his name to Henry Wilson and asserted that he was a naturalised Hollander, thus hoping to escape recognition when he landed in England. Some of his fellow prisoners who did not know when they disembarked at Fowey in Cornwall, of the proclamation offering a reward for his arrest, promptly sought out a local Justice of the Peace, to whom they denounced Crosfield for his seditious talk, his singing of revolutionary songs, and his boasting that he had shot at His Majesty but had unluckily missed him.

Crosfield was arrested in August 1795 and committed to prison by the magistrate. As he was being taken in a coach across the moors to Bodmin prison, he tried to escape; he offered his two guards a guinea each if they would take the handcuffs off his hands and, being refused, he raised the offer to two guineas each. When one of his guards asked him what he would do with the driver of the coach, he replied, in the guard's own words: "If I would let him have one of the pistols, he would pop at him and soon settle that business." A few days later he was conveyed to London, imprisoned in the Tower and charged with high treason.

The trial of the four men previously charged with high treason had been postponed because Upton had jumped his bail and could not be found. The Attorney-General was now compelled to proceed with the prosecution of Smith, Higgins, Le Maitre, and Crosfield, but not collectively. After nearly eight months in the Tower, Crosfield was put in the dock.

This was the fifth trial on charges of high treason since the suspension of Habeas Corpus in 1794. The previous four had failed. The Attorney-General [Sir John Scott (1751–1838), later Lord Eldon, Lord High Chancellor] had himself appeared for the prosecution at the trials of Thomas Hardy (1752–1832), John Horne Tooke, and John Thelwall, the leaders of the London Corresponding Society, and had been unsuccessful. It was therefore with reluctance that he undertook the task of appearing for the King versus Dr. Crosfield. He must have remembered that when he left the Old Bailey every evening at the previous trials he had been hissed by the mobs who assembled outside.

The indictment against Crosfield was that he "did compass, imagine, and intend to bring and put our lord the King to death; that on the 1st September in the parish of St. Dunstan's procured, made, and provided a certain instrument for the discharge of an arrow, to be loaded with poison; that he caused to be employed John Hill to make two pieces of wood to be used as models; and that he caused to be employed one Thomas Upton to assist in making the said instrument."

The most eminent advocates of the period, led by the Attorney-General and Solicitor-General appeared for the Crown; Mr. Adam and John Gurney (1768–1845) defended the prisoner, and the trial began on May 11, 1796.

Upton, the chief witness for the prosecution, could not be found. His wife, Elizabeth Upton, said in the witness box that she last saw her husband at their home in Wapping on February 22, when he left the house at eight o'clock in the evening and, before leaving, had given her his seal. The following morning a waterman brought her a hat which she identified as belonging to her husband, believing that he was drowned.

Mr. Adam, for the defence, pleaded, "It was strange that a person should be drowned and his body never found; not anything belonging to him but his hat, especially in this country, where so much pains were taken to investigate a circumstance of that nature. Whether Mr. Upton was to appear again in this world he would not pretend to say; but he would certainly lay before the Jury as strong evidence of his being in life as Mrs. Upton had given of his death." Here the Attorney-General signified that he had objections to evidence of that nature.

Mr. Adam said that if he could raise a doubt that this man was not dead, but gone out of the way, he would raise a presumption that the story of the seal was continued for the purpose of imposing the opinion that he had committed premeditated suicide—that it was part of a trick and plan proceeding from a diabolical mind. He was far from any intention of imputing the absence of this person to the Gentleman who conducted the prosecution. He judged it, however, absolutely necessary to bring evidence of its existence before the jury.

The prisoner, previous to his trial, had been brought before the Privy Council and interrogated, but in view of his impending trial had wisely refused to answer questions. He did, however, say regarding his flight to sea, that "he hadn't a guinea, so meant to go to some country where he could exercise his talent for Natural History."

Messrs. Dowding, Bland, and Flint, the three brass-founders, and Hill, the wood-turner, who had in turn been visited by Crosfield, Upton, and Palmer, all recollected Upton coming to their shops because he was most noticeably lame on account of his club foot. They unanimously agreed that Upton had made all the enquiries, negotiations as to cost of the articles, etc. They said the other man merely explained the drawings but none of them identified this man as the prisoner. Only one of the craftsmen identified Peregrine Palmer, and indeed it was only Palmer himself, as a witness for the prosecution, who was able to assert that Crosfield was a member of the party.

A mathematical instrument maker, David Cuthbert, was the possessor of an air-gun in which Upton had taken a more than ordinary interest. On one occasion Upton had come to see the gun, accompanied by a "gentleman." Cuthbert told the court that this gentleman said that he "was very fond of shooting and that he had lost some of his fingers by the explosion of a gun." He did not, however, look at the gentleman's hands. Upton had declared his companion was Crosfield, but the witness could not identify him.

Several of Crosfield's fellow prisoners of war were Crown witnesses. John Le Breton, a boat-steerer on the *Pomona*, said that Crosfield told him he had invented an air-gun with which to shoot the King, and also an arrow to be fired from it: His Majesty was to be shot at in the playhouse. Le Breton added that Crosfield drank a lot, was always very talkative, but he never knew whether he was in "jest or earnest."

James Winter, a master mariner, said that Crosfield boasted he was Tom Paine, that he had had a pop at his Majesty between the Palace and Buckingham house, but unfortunately missed him, and that he "hoped to live to see the day when the streets of London should be up to his ankles in the blood of the King and his party." The ex-prisoners of war who appeared as Crown witnesses all agreed that Crosfield's conversation was of a seditious nature, that he was always singing seditious songs, one of which was, "Damnation to the King," and another "Plant, Plant the Tree." His general behaviour was obnoxious and most offensive, and they strongly resented his revolutionary attitude.

Captain Anthony Collins, who commanded the prison ship at Brest, spoke favourably of Crosfield who, he said, had come aboard his ship voluntarily to take charge of an outbreak of sickness which had affected a large number of the prisoners. "I am confident," said Collins, "that he saved fifty or sixty lives by his great care and attention."

Further evidence of character for the defence was given by two surgeons, one of whom said that he was "a man of humanity; of good character," whilst the other said that he was "an exceedingly good man, incapable of committing any crime." Three former landladies, with each of whom Crosfield had lodged for several years, said they had nothing but good to say of him, as he was a good-natured man who would hurt nobody; in fact his fault was that he was too good-natured.

W.H. Pusey and George Steers, fellow clerks with an attorney, declared that Upton had shown them a brass tube at a divisional meeting of the London Corresponding Society which resembled the brass tube exhibited in court. Upton had sworn to the Privy Council that the tube had been shown to the witnesses by John Smith.

The defence sought to establish a motive for Upton's sudden determination to denounce his fellow members as having conspired to assassinate the King. He had falsely sworn as to the method by which this dastardly act was to be accomplished and he himself had devised the deadly weapon.

James Parkinson was the first witness to appear for the defence of Upton and was in a position to provide the jury with the ulterior motive for Upton's machinations. He described himself as a surgeon and apothecary, residing at Hoxton Square. In the examination-in-chief he said he had been a member of the Committee of Correspondence of the London Corresponding Society in the months of August and September 1794, and that Smith and Higgins were also members of that committee but not Le Maitre:

Mr. Gurney: Were there to your knowledge, any inquiries instituted in that committee or in the general committee, by either Le Maitre, Higgins, or Smith, respecting Upton?

Parkinson: By Higgins and Smith there was an inquiry instituted, at the request of the Committee of Correspondence, among whom I was one, who was very solicitous for that inquiry into the character of Upton.

Mr. Gurney: What was the charge that was brought against Upton?

Parkinson: It was stated either by Mr. Smith or Mr. Hodgson, that he had heard it reported that Upton had set his house in Cold Bath Fields on fire.

Mr. Gurney: Was that inquiry pursued to any considerable length?

Parkinson: They were desired at a meeting of the committee to make the inquiry.

The Lord Chief Justice, Sir James Eyre, who was the presiding Judge of the court, intervened at this stage and interrupted Mr. Gurney by saying, "If you go to the point to show that there was any emnity between Smith and Higgins and Upton, very well," but his Lordship did not want to hear all the detailed proceedings of the committee.

Parkinson: I can only speak of any dispute that subsisted between them by the report of Smith and Higgins.

Mr. Gurney: That is not evidence. Did you upon that occasion see Upton yourself?

Parkinson: Only once; which was for the purpose of carrying a letter to be delivered to him expelling him from the society.

In the report of the trial Parkinson is quoted as having said concerning the dispute, "A paper signed Pasquin sometime afterwards appeared, which was a virulent satire on the individual members of the general committee, a false and unsupported accusation against their members. Of this letter Upton avowed himself the author, which produced a motion of censure, and expulsion against him, which was made by Le Maitre and supported by Higgins."

Upton was furious at this and directed his anger against Le Maitre for proposing his censure, and against Higgins, the seconder of the motion, who had remarked that the motion would have to be carried quickly otherwise Upton "would be hopping it off." Upton took this to be directed against his club foot. "Higgins," he said, "was mocking him." Higgins turned on Upton, whom he called a "liar" and an "immoral blackguard," for which remarks Higgins was called to order and made to apologise. When the motion of censure was put to the meeting and passed, Upton flew into an uncontrollable rage, the meeting was thrown into disorder, and it took 20 members to pacify him.

Just at this time subscriptions were being solicited for a fund to provide financial assistance for the families of members who had been imprisoned. Upton was an authorized collector, but the committee now felt obliged to withdraw his name from the printed list of receivers of donations. Parkinson, by order of the committee, wrote a letter notifying Upton of this decision. The next thing that the members heard was that Smith, Higgins, and Le Maitre had been arrested and charged with a conspiracy to assassinate the King. Rumours had spread that Parkinson was among those arrested, and he wrote, "for, in consequence of the newspapers having declared me to have been implicated with you (Smith), in the horrible plot, a near relation actually received a condolatory visit from a friend, who had heard, from various persons, that I was in DOUBLE IRONS in Newgate, in consequence of the charge against us having been completely made out."

Mr. Adam stressed that the identify of Crosfield was established by only one witness for the Crown, Mr. Palmer. The jury was reminded that the brass tube was found in Upton's possession and that there was absolutely no evidence that it was ever in Crosfield's possession. All Crosfield did was to explain the drawings.

"If Crosfield," said Mr. Adam, "had been the person who conspired, why not apprehend him when the others were apprehended, but it was so long as the 27th February 1795, before the Crown officers thought proper to issue a proclamation offering a reward . . . .The learned counsel laid great stress on Mr. Crosfield's refusal of the situation offered him as Surgeon to the Naval Hospital at Brest, and voluntarily returning to this country."

After Adam's summation, the Lord Chief Justice asked Crosfield if he had anything to say. "My Lord," replied Crosfield, "I have nothing whatever to add to what was stated by my counsel; however I may have been represented, I am fully confident of my own innocence, and firmly rely on the justice and humanity of a Jury of my own countrymen." The jury, after an absence of two hours and twenty minutes, returned a verdict of not guilty. It was a popular verdict, "the people in the court, among whom the prisoner appeared to have many friends, demeaned themselves with the greatest degree of decorum, but no sooner was the result communicated out of doors than the air was rent by the huzzas and acclamations of the populace."

While imprisoned in the tower, Crosfield had written *Remarks on the Scurvy* (Crosfield, 1797). After his release he dedicated it to the members of the jury, and wrote in the preface that Upton was "insane" and "actuated by malice" (Crosfield, 1797, Preface).

The trial of Smith, Higgins, and Le Maitre took place at the Old Bailey a week later, on May 19, 1798, and they were placed in the dock together. Previous to this they had been examined before the Privy Council, where Parkinson had been a witness on their behalf. This investigation had proved inconclusive. The Attorney-General offered no evidence against them (nolle prosequi) and the members of the jury were immediately directed to find the prisoners not guilty. So ended the pop-gun plot which had existed only in the imagination of one man, Upton, but which had placed four men in peril of being hanged, drawn, and quartered, the penalty for high treason, which had never been repealed. It was clear that Upton was the villain in this drama, and that it was his cleverly designed scheme into which he had unveigled Crosfield, who in his cups was weak and easily led, perhaps merely to make some easy money. The good-natured Crosfield had known Upton for only one month.

After his acquittal Crosfield resumed his revolutionary activities, working to revive the almost defunct London Corresponding Society, and in 1797 was elected its President. Trouble soon followed. An address was sent by the London Corresponding Society to the Society of United Irishmen, which began, "Brave and suffering nation," and was signed R.T. Crosfield, President, Thomas Evans, Secretary, and dated January 30, 1798.

An insurrection in Ireland had been planned by the United Irishmen, which was to take place on March 23, 1798, but the Government had been forewarned. The address signed by Crosfield, which contained many seditious and treasonable passages, was shown to a Committee of Secrecy set up by the House of Commons. The arrest of Thomas Evans, the Secretary, a bookseller of 46 Paternoster Row, followed. When Evans was searched after his arrest, a copy of an "oath" administered to would-be members was found in his pocket, contrary to an Act of Parliament which expressly forbade this. On February 7, 1798, a writ of Habeas Corpus was issued to bring Thomas Evans to court, charged with high treason. It seems strange that Crosfield was not arrested at the same time as Evans, but again the bird had flown and could not be found.

So, on March 14, 1798 a King's Proclamation was issued which informed the public of a reward of £200 for apprehending Robert Thomas Crosfield, "whereas Robert Thomas Crosfield, late of Cursitor Street, Chancery Lane, surgeon, stands charged with suspicion of High Treason."

A few months later Crosfield was arrested. In the Morning Chronicle of August 6, 1798 is the following account of arrest: "A paper of yesterday states that on Saturday, Mr. R.T. Crosfield, for whose apprehension a reward of £200 was some time since offered by the Government, was taken to the Police Office in Worship Street, Shoreditch, by two soldiers of the East London Militia. He was removed to the Duke of Portland's Office, where he underwent a short examination before Mr. Wickham and Mr. Flood, and committed from thence to Tothill Fields, Bridewell. He is charged with High Treason. About a week since he took lodgings in Goswell Place; a female who inhabited the adjoining house had frequently met him in the chambers of a Gentleman in the Temple, and on Saturday evening seeing him in his garden, recognised him and gave information to the soldiers who apprehended him."

The Government was alarmed by the insurrection in Ireland promoted by the United Irishmen's Society, and still more frightened by the formation of similar societies in London. The Committee of Secrecy had reported, "that one of the plans proposed (the beginning of last year) was that of a general insurrection . . . the objects of which were to seize or assassinate His Majesty, the Royal Family, and many of the members of both Houses of Parliament. They were to be headed by a person of some experience in His Majesty's service: but the apprehension that their force was inadequate for such an attempt deterred them in its execution."

On April 20, 1798, the members of the executive committee of the London Corresponding Society were all arrested when they met in Wych Street, Strand; among them was Paul Thomas Le Maitre, "in custody for treasonable practices." The Government acted swiftly. On or about April 27, 1798, a bill was introduced in the House of Commons for the more effectual suppression of societies established for seditious and treasonable purposes, which included all societies of United Irishmen, United Scotsmen, United Englishmen, as well as the London Corresponding Society, and all societies whose members were required to take an oath. Printing presses were to be licensed and all publications were to bear the name and address of the printer. Rooms used for debates or lectures were also to come under supervision.

The State Trials of Mandy, Tooke, and Thelwall, in 1794, and of Dr. Crosfield, in 1796, had made the Government cautious. Had they been indicted for "misdemeanour or sedition only," conviction would have been certain. In spite of the indictments of high treason against Crosfield, Evans, and Le Maitre, they were not brought to trial for high treason. The Habeas Corpus Act was suspended and the men detained in custody for an indefinite period, until the societies were finally suppressed and the danger caused by their subversive activities had passed. This time the prisoners were well treated, sent separately to country prisons such as Reading and Winchester, where they were "allowed reasonable subsistence, but were strictly forbidden all correspondence, out of the prison, and all intercourse within." All were released the following year.

Of the four men, Smith, Higgins, Le Matire, and Crosfield, only Higgins did not continue with a rebellious attitude toward the Government. Smith, who was a printer and bookseller, was, in 1797, given two years hard labour and fined £100 for publishing *A Summary of the Duties of Citizenship, the tendency of which is to undermine and destroy all law, all government and all religion* (Anon, 1797).

Crosfield's illness was not improved by a second period in prison, and his health was slowly degenerating. Anticipating his end he retired to Hendon, Middlesex, where he died after two years. His tombstone, which has been well preserved by the parish, is still seen near the pathway on the east side of the church. The inscription included his self-composed epitaph, in which he summed up his own philosophy of life:

In memory of Robert Thomas Crosfield, M.B., son of the late Francis and Ruth Crosfield of Spennithorn in the County of York who died the 8th day of November 1802, aged 44 years. Previous to which he wrote the following Epitaph so truly characteristic of himself

Beneath this stone Tom Crosfield lies
Who cares not now who laughs or cries
He laughed when sober but when mellow
Was a Harum Scarum heedless fellow
He gave to none designed offence
So Honi Soit qui Mal y Pense

With this outline of the events before us, we can now examine Parkinson's role in the Pop-Gun Plot in greater detail.

In the preface to his pamphlet, *Assassination of the King* (Smith, 1795), John Smith describes Parkinson as "that worthy citizen who did not scruple to come forward in the midst of danger, and assert before the Privy Council, the innocence of the supposed conspirators" (Smith, 1795, Preface). He was courageous and confident in his resolution to risk voluntary involvement in the so-called Pop-Gun Plot, for at this time he had to consider his young family and his large medical practice. An appraisal of the official account of the secret enquiry shows that Parkinson's frankness, sincerity, and integrity made a great impression on their Lordships in spite of the fact that he had come before them as a committeeman, a direct successor of leaders of the distrusted and dreaded London Corresponding Society.

Smith had been confined in Newgate to a room totally devoid of furniture which had unglazed windows stuffed with paper, during one of the worst winters known, with the result that he became desperately ill and received neither nursing care nor medical treatment. "A damp room, and cruel treatment," wrote Smith in a letter to Parkinson, "have so injured my constitution that I have been confined to bed for some time dangerously ill. I beg you will come and see me."

Parkinson was given a permit by the Privy Council to visit Smith, after which he wrote a respectful plea to the Council, saying that Smith, "had symptoms of disease of the lungs," and that "he spat a considerable quantity of matter, of such an appearance, as to render it dubious whether an ulceration of the lungs might not already have taken place; the accompanying symptoms were such as to convince me that his longer continuance without the requisite care and attention could hardly fail to *occasion his death*." The Privy Council, in response to Parkinson's request, ordered that Smith should be removed from the Felon's side to the State side of the Newgate Prison and provided with a physician, whilst Parkinson was permitted to see him at any time.

On April 27, 1795, Parkinson was asked to call upon Mr. Richard Ford, Secretary to the Privy Council. At this interview Parkinson was informed that, since Crosfield could not be found, Mr. Pitt had decided that Smith, Higgins, and Le Matire would be released on bail. Parkinson remarked that it was strange that when Crosfield was available no attempt was made to arrest him but that as soon as the Government knew he was safely out of the country, a reward was offered for his capture. Mr. Ford requested Parkinson to ask Smith to write to the Privy Council formally asking for bail, but at the same time advising Smith not to use "strong language," or the "language of rudeness" in his letter. Parkinson assured Mr. Ford that Smith was not the type who would send a begging letter, and that from his knowledge Smith was not the kind of man who would wish for bail unless his fellow prisoners were granted the same facilities. Parkinson then desired to know about Upton, if he was to be released on bail on the same terms, and ventured the opinion that if he were in Smith's position he would choose to remain in prison if Upton, his accuser, was also released. Mr. Ford replied that the Privy Council were in a quandary as to what to do with regard to Upton.

The opportunity of questioning Mr. Ford was one not to be missed, and Parkinson took advantage of the fact that he could at least talk freely to a spokesman for the Government. He asked Mr. Ford for information concerning the letters from Smith, Higgins, and Le Maitre, which were exhibited to the Council, and which the accused men alleged to be forgeries. Mr. Ford thereupon produced a letter the Privy Council had received from Le Matire, in prison, and for comparison showed the letter that Upton alleged he had received from Le Matire. Although Mr. Ford did not express the views of the Council, or his personal opinion, it left Parkinson in no doubt that they all thought it was a forgery.

Parkinson had refused to take the oath at the Privy Council's enquiry until he had an assurance that he would be questioned only concerning Smith, Higgins, Le Maitre, and Upton, and their connection with the pretended plot and nothing else.

Their Lordships were at first exceedingly angry with his refusal to take the oath. The Attorney-General, who was to put most of the questions to the witness, then intervened and said, "You will not be asked to criminate yourself." To which Parkinson replied, "There is no question you can produce an answer to criminate me."

The Attorney-General, however, in spite of their Lordship's assurance that no irrelevant questions would be introduced, did ask irrelevant questions, which put the examinant on his guard. Richard Hodgson, a hatter of Broad Street, Westminster, a member of the Committee of the London Corresponding Society, had earlier been charged with high treason in common with 11 other members of the committee, amongst whom were Hardy, Horne, Tooke, and Thelwall, but had not been arrested like the others because he could not be found. Suddenly, Parkinson was asked if he had seen Hodgson, and when and where he had seen him, the Attorney-General thus hoping to obtain information which would lead to his arrest. Parkinson was in an uncomfortable and dangerous situation. He declined to say *where* he had seen Hodgson, though he was forced to admit having seen him the previous night.

Mr. Pitt: What do you mean, Sir? You must answer the question?

Parkinson: This question, any more than many others I have already answered, can have no reference to the pretended plot.

Mr. Pitt: That does not signify; you are bound by the oath you have taken to answer it.

The Secretary was then asked to repeat the oath, which he read but did not ask the witness to reaffirm.

Parkinson: But it was surely agreed that I was to object to certain questions.

Mr. Attorney-General: (Pitt's interjections deleted) Yes, to such as might criminate yourself, and to those only. . . . Choose whether you will answer the question or take the consequences of refusal . . . . Now, Sir, where did you see Hodgson?

Parkinson: I wish not to behave with incivility towards your Lordships, but I must say, and that not without considering my answer, that I am used exceeding ill.

. . . .

Mr. Pitt: Sir, you cannot object to this question.

Parkinson: I conceive that I *can*, and *do* on this ground also. That you ought not to put such questions, the refusing to answer which will imply crimination.

Mr. Attorney-General: Sir, you must answer the question.

One of their Lordships: Unless it will criminate yourself.

Parkinson: My Lords, my legal knowledge is but very trifling, from the extraordinary circumstances I have lately observed, I know not what may now be decided crime or not. On that ground also I object to answering this question.

Mr. Attorney-General: Then, Sir, you know that a bill was found yesterday for high treason against Hodgson, by a Grand Jury of his countrymen?

Parkinson: I learned so from the public papers, and indeed, from himself.

Mr. Attorney-General: Now, Sir, if after that you conceal him, you are guilty of misprison of treason.

Parkinson: That, Sir, I am aware of.

Mr. Attorney-General: Now, Sir, answer the question; but you are not wished to criminate yourself.

Parkinson: Then, I saw him in my own house.

(Sequence of exchanges about Hodgson's whereabouts ending with Pitt writing on a slip of paper.)

Attorney-General: Do you know where he is now?

Parkinson: No.

Attorney-General: Do you know where we can find him?

Parkinson: No.

Having established the fact that Parkinson need not give replies which would incriminate himself, the Attorney-General proceeded to ask questions which Parkinson considered were irrelevant to the matter in hand. The Privy Council, through Government spies and informers, had heard of seditious songs (e.g., 'Plant, plant the tree') and toasts and was anxious to obtain confirmation of this. Parkinson said he had attended the last dinner of the Constitutional Society.

Attorney General: Do you remember the toast *Success to the armies fighting in the cause of liberty?*

Parkinson: I remember that perfectly well.

Attorney-General: How did you understand that toast?

Parkinson: I understood it to be *equivocal.* Since the well wishers to the French would I suppose contend it meant their armies; whilst the friends of the combined armies would claim it otherwise.

Some 20 toasts in all were drunk at the dinner, and it was fortunate for Parkinson that he was not asked for an opinion of other and more seditious toasts that were drunk at the same dinner. He was shown a political pamphlet, entitled *A Vindication of the London Corresponding Society*, which had been published anonymously. The Lords were most anxious to discover the name of the author.

Attorney-General: Pray Sir, did you ever see one of these pamphlets?

Parkinson: Yes, my Lords.

Attorney-General: Do you know the author?

Parkinson: Exceedingly well.

Attorney-General:	Who is the author?
Parkinson:	I object to the question.

One of their Lordships then said to the clerk, "Put down—Refused to answer *because* it will criminate himself." Parkinson, in anger, struck his fist on the table, and declared, "My Lords, I claim that no words be inserted in my examination as my answers but what proceeds from my own lips. I gave no such reason. (Another of their Lordships to Fawkner.) The question itself, my Lords, is not itself worth contention.

Parkinson:	If put again I will answer it.
Attorney-General:	Who is the author?
Parkinson:	I am the author.
Attorney-General:	You are the author?
Parkinson:	I am the author.

When the 11 members of the Committee of Correspondence of the London Corresponding Society were confined to the Tower, Hodgson, the only one still at large, immediately formed a new committee, which included Parkinson.

Mr. Pitt:	Pray, Mr. Parkinson, how come you to be invited on this Committee?
Parkinson:	Because, I believe they did me the honour to believe me *firm* in the cause of parliamentary reform. (Here, said Parkinson, *Mr. Pitt smiled* when he heard this reply, and it must indeed have given Parkinson some satisfaction to have been able to make a slight hint at Mr. Pitt's recantation of his former staunch support of parliamentary reform). And (continued Parkinson) because I had just published a little tract . . . (for the benefit of the wives and children of those imprisoned for high treason)
Attorney-General:	Pray what was that?
Parkinson:	*Revolutions without bloodshed.*
Attorney-General:	That was yours too?
Parkinson:	It was.

He was shown a handbill which was a burlesque imitation of a playbill, entitled *La Guillotine, or George's Head in a Basket* which he acknowledged he had seen so long before that he did not remember how he obtained it.

Attorney-General:	What do you think of it Mr. Parkinson?
Parkinson:	Am I to give my opinion too?
Attorney-General:	Yes, Sir.
Parkinson:	Why, I think it is a silly—ridiculous—and I may say impudent thing.

The Privy Council was intrigued by the Committee of Correspondence, the Committee of "Secrecy" as it had been called by Upton, because the names of its members were known only to the members of the committee themselves, and not to members of the Society generally. In the view of the Council the so-called Secret Committee had something sinister about its deliberations and intentions. Now that they had one of the members of this committee in front of them they were determined to take this opportunity to find out more about its purpose. Parkinson's replies were cautious and evasive.

Attorney-General:	You were invited to the *Committee of Correspondence*?
Parkinson:	Yes.
Attorney-General:	It was called the *Secret Committee*, I believe?
Parkinson:	I never heard it described as such. *Upton* once employed the term, as a term of reproach and was reproved for it.
Attorney-General:	What were the functions of this committee?
Parkinson:	I do not know.
Attorney-General:	Not know—that is very strange, is it not?
Parkinson:	No—I never heard them defined.
Attorney-General:	So, after seeing this handbill, you could go into a Committee, the functions of which you knew nothing about?
Parkinson:	I said, my Lords, I believe I had not seen the handbill. (This was a reference to another handbill shown to Parkinson, the nature of which was not disclosed in the official report, and he had said, "I rather think, my Lords, I never saw it before." The Secretary then intervened by reading Parkinson's previous answer, denying that he had ever before seen that particular handbill.)
Attorney-General:	Well, well. You could go into a committee without informing yourself of its functions?
Parkinson:	I did, because I conceived myself not confined to it. I could have withdrawn myself if anything had passed in the Committee that I did not approve; but I saw nothing of that kind. I am therefore a member at the present moment.

Parkinson, it seems clear, had proved that he could not be drawn into giving the Council the information they so eagerly sought regarding the objects of the so-called Secret Committee. He parried so successfully with the Attorney-General that it is not to be wondered that Sir John did not forget Parkinson or his evidence given that day.

It was a sore point with the Government that it had failed to get possession of the minutes of the Committee of Secrecy after a thorough search of the residence of the members, but the Attorney-General succeeded in obtaining from Parkinson the name of the new Secretary and what had happened to the minutes.

Attorney-General:	Who was the Secretary?
Parkinson:	I hardly know; but I believe Burks was the appointed Secretary.
Attorney-General:	Do you not know that Smith was the real Secretary, and Burks only the nominal one?
Parkinson:	I do not.

Attorney-General:	Who kept the papers?
Parkinson:	Mr. Beck.

(Question from Scott as to Beck's whereabouts. He was last seen at Parkinson's house.)

Attorney-General:	Where are those papers now?
Parkinson:	Beck told me with regret, and with regret I heard it, that they were burnt.
Several of their Lordships:	Burnt! Burnt!
Parkinson:	Yes, my Lords.

John Beck, a saddler of Oxford Street, had anticipated a search of his house, and had asked his servant to take the papers away from the house out of reach of the Government searchers. The servant, however, found it impossible to escape with the papers as the searchers were already inside the house, so he burnt them. Parkinson, who gave this information to the Council, said the servant had "foolishly burnt them."

Parkinson's political creed was confined to agitating for a reform of parliamentary representation, a system of universal suffrage, and a yearly renewal of the House of Commons. He left the London Corresponding Society when it got into the hands of extreme republicans who attempted to enlist the support of the French Government to establish a republic in this country, and who were willing to use force for the furtherance of their intentions.

Although the Pop-Gun Plot may now be considered as something of a storm in a teacup, it was without doubt an important trial. Had the prosecution of Crosfield been successful, the precedent would have threatened the lives, or at least the liberty, of a great number of people whose actual "crime" may have consisted only in disagreeing with the government of the day. Parkinson showed himself to be a man of principle and courage by defending those whom he believed to be innocent.

CHAPTER 5

The Popular Medical Writer

Medical Admonitions was the title of a book on domestic medicine written by James Parkinson and first published in 1799. An octavo-sized book divided into two parts which are described as volumes, it contains some 548 pages.

Volume one begins with an *Address to the Reader* in which the author says: "Convinced that many lives are lost by neglecting to apply sufficiently soon for medical aid, and by improper treatment of disease by domestic practitioners, the writer of the following observations offers them to the public, with the pleasing hope that they may lessen the number of victims to negligence and presumption."

An *Introduction to the Table of Symptoms* begins:

In the most trifling affairs, hardly anyone will be found who will entrust another, with the performance of any business of the nature of which he is uninformed.

Should anyone be asked to trust another with the turning of a tooth pick, out of some substance which he valued, merely on the recommendation of his having an excellent set of tools, and an essay in the art of turning, he would consider that some practice would be necessary to give him the art of turning the wheel, with due velocity and regularity, and of holding the tool, and of applying it in the most proper manner. He would remark that repeated observation and experience were required, to give a knowledge of the grain of various substances, and in what direction the grain of different substances should be worked; and that without this knowledge, he would be likely to shiver to pieces the substance on which he operated, instead of making an useful and beautiful form. He would therefore say, No, you must excuse me, I must apply to some artist of more judgement and ability, for I will not have my tooth pick spoiled. But often when life itself is at stake, much less circumspection is employed than in the turning of a tooth pick; the treatment of some dangerous disease is confided to anyone who possesses a medicine chest and the small share of skill which is derived from the perusal of some treatise on domestic medicine.

To destroy the strange infatuation on which such inconsistency depends, would be a hopeless endeavour; but humanity suggests, nay, commands, that everything be done which may be likely to lessen the evils resulting from it. If men will, in spite of remonstrance rush into danger, they are not, therefore to perish unaided. Should anyone obstinately put to sea without a compass to steer by, and without any knowledge respecting the navigating of a ship; but what he picks up during his voyage, by reference to some treatise on navigation, it would not be sufficient, merely to endeavour to dissuade him from making the rash attempt. But if he persist, every possible assistance should be yielded him: the perils he has to shun should be clearly pointed out, the different rocks and quicksands he is to avoid should be marked; and the different circumstances should be described which may show his near approach to danger.

With a similar intention, is the following table: it is by an attention to symptoms, that the physician is enabled to steer his course with confidence and safety and to discover the dangers which threaten. By a reference to the following table, and by examining the symptoms most prominent at the commencement of disease, domestic practitioners therefore, may more readily detect the disease which they wish to remove; and will be then better able to determine, as to the propriety of taking the task of curing it on themselves. If this be resolved on, a reference to the table may also serve to show the degree which is marked by any particular symptom, which may arise in the progress of the disorder, and which calls for very powerful assistance (Parkinson, 1801, Introduction).

Parkinson's Table contains an alphabetical list of symptoms, beginning with anxiety; appetite (loss of); and ending with wakefulness; weakness; yawning. The remainder of the first volume enumerates various diseases, with a description of each, its symptoms and its treatment. Prevention is not overlooked, and the dangers and difficulty of recognition of a disease from its first symptoms is stressed. What is today termed a physical sign, was, in those days, included under the term symptom.

Dr. Buchan's (1729–1805) book on *Domestic Medicine* (Buchan, 1769), was in the latter part of the 18th century an indispensable requirement in almost every household, but Parkinson criticized some of Dr. Buchan's remedies. "I shall occasionally point out some passages in your favourite book (Buchan's *Domestic Medicine*) as seem to be more productive of harm than of good" (Parkinson, 1801, p. 20).

"Dr. James's Powder, and Emetic Tartar" were much advertised nostrums which Parkinson attacked as being "so violent as to require the utmost caution in administering them."

The celebrated Dr. Goldsmith lost his life by the imprudent administration of Dr. James's Powder, as was clearly proved in the account of his illness, published by his worthy medical attendant, Dr. Hawes (Parkinson, 1801, p. 44).

The author advocated fumigation of rooms by "nitrous or marine acid in a state of vapor," a method then lately introduced by Dr. James Carmichael Smyth (1741–1821), to be used "to lessen the probability of the contagion being disseminated, and for correcting the air of the apartments" in cases of contagious fever. "Another caution not in general sufficiently attended to," the author wrote, "is to properly ventilate, and fumigate the clothes, which had been worn by the patient" which "may still retain a sufficient quantity of the poison to communicate the disease" (Parkinson, 1801, p. 65:67).

Parkinson was a strong believer in the efficacy of fresh air in the treatment of putrid fever (typhus fever), and quotes a personal experience of his friend, the Rev. Joseph Townsend (1739–1816) of Bath, a fellow geologist.

> I have given a slight sketch of the principles on which airs of different purity are employed in the different diseases, whilst treating of Consumption. I shall therefore here only introduce the evidence of the ingenious Rev. Joseph Townsend, the benevolent author of the Guide to Health, etc. as to the use of vital air in this disease. "In my putrid fever, whenever the air of my chamber was artificially *oxygenated*, as my physician Dr. Thornton, often witnessed, my respiration was pleasant, my oppression at my chest relieved, and I was enabled to breathe freely through the nostrils, without the assistance of my mouth, which I could not do before the room was oxygenated" (Parkinson, 1801, p. 70).

On the subject of teething, lancing of the gum is strongly recommended, despite the assertion of Dr. Buchan that it was "seldom known to be of great benefit," in cases where convulsions have resulted.

> So little pain is occasioned by lancing the gum over the distending tooth, and such immediate relief is often derived from it, that it frequently happens that a child who, the moment before, appeared to be in great agony, will bestow a grateful smile, the instant the incision is made; and if, as sometimes happens, the gum grows again over the tooth, will not only cheerfully submit to the operation, but even solicit it. I have been most earnest in urging parents to permit this little operation . . . (Parkinson, 1801, p. 147).

He also remarks that lancing the gum was upheld by John Hunter (1728–1793).

Of pulmonary consumption or ulcer of the lungs, the author observed that "no period of life, from childhood to old age itself, is exempt from this disease; but the most frequent subjects of it are the young and the middle aged." His views on atmospheric air are interesting: "atmospheric air," he wrote, "contains two different airs: the one termed *vital* or *oxygen* air, which supports respiration and combustion; the other called *azotic* air, in which combustion soon ceases, and animals quickly die." "Oxygenated air," he proceeds, "has been successfully employed in putrid fever, asthma, dyspnoea," by Drs. Beddoes, Thornton, Carmichael, Smyth, Darwin, Ferriars, Pearson, etc., and adds that "*azotic* air, and chiefly air termed the *hydrocarbonate*, has been successfully applied by the same gentlemen in pleurisy, croup, spitting of blood, and even in consumption." "Hence then it appears, that so far from a removal into the pure air of the country being indicated in every case of consumption; that cases must often occur (those whose inflammation persists) which the air, even of a crowded city, may be preferable" (Parkinson, 1801, pp. 171; 182–3; 184).

In cases of apparent death from drowning, suffocation, cold, etc., Parkinson recommended the "resuscitative process" formulated by the Royal Humane Society, enjoining the reader that "What thou doest, do quickly." "The directions given by the Royal Humane Society," he counsels, "are so clear and explicit, as to put it in the power of anyone, to employ the most powerful means for the recovery of the apparent dead. As no opportunity should be permitted of promoting the diffusion of knowledge so important to the dearest interests of humanity, I have taken the liberty to introduce the process recommended by the society" (Parkinson, 1801, pp. 389; 388).

The instructions are given in detail and include cleaning the mouth and nostrils; laying the patient near a fire; gently rubbing the body with flannel sprinkled with spirits or flour or mustard. "To restore breathing," it is advised, "introduce the pipe of a bellows (when no apparatus is at hand), into one nostril; the other and the mouth being closed, *inflate the lungs*, till the breast be a little raised; the mouth and nostrils must then be let free. Repeat this process till life appears. *Tobacco smoke* is to be thrown gently into the fundament [anus], with a proper instrument, or the bowl of a pipe covered so as to defend the mouth of the assistant." "The plans of resuscitation," it is suggested, "to be used for three or four hours. *It is an absurd and vulgar opinion, to suppose persons irrecoverable, because life does not soon make its appearance.*" Finally, it is stated that bleeding is never to be undertaken except by a medical assistant (of the society presumably), who is also requested to use "electricity" (Parkinson, 1801, p. 390; 392).

Parkinson quotes as follows from his *Account of the Effects of Lightning*:

> That the electric fluid acts as a stimulus upon the animal system, must, I believe, be allowed; but that its stimulating effects are more transient, and the succeeding state of debility consequently more rapidly produced, than what results from the application of any other known stimulus, must, I think, be also allowed (Parkinson, 1801, p. 380–1).

Electrical treatment was the latest remedy. Only a few years previous to the publication of *Medical Admonitions*, the London Electrical Dispensary had been founded, in 1793, at No. 16 Bunhill Row, only a few minutes from Hoxton Square. Its object was "the administration of Medical Electricity to the Poor, for such complaints to which its application may be useful."

In the section dealing with "Worms," the author states:

> The necessity of determining, with some degree of accuracy, whether the complaints of a child depend on worms, or not, is much greater than may at first appear (Parkinson, 1801, p. 402).

Again, he returns to the subject of quack remedies and their danger, quoting an interesting case.

Concerning the treatment of "Weakness of the Stomach," the author observes: "There are few diseases, if any, over which the physician has so little power; unless aided by the resolute efforts of the patient, as in this." It was difficult to persuade patients that their stomach troubles were due to overeating or overdrinking.

Like others of his time, Parkinson thought about "Hypochondriac Affection." He defined the malady as:

Want of resolution and activity with respect to all undertakings; a disposition to seriousness, sadness, and timidity; as to all future events, an apprehension of the worst, or most unhappy state of them; and therefore, often upon slight grounds, an apprehension of great evil (Parkinson, 1801, p. 413).

He continues:

Such persons are particularly attentive to the state of their own health, to even the smallest change of feeling in their bodies; and from any unusual feeling, perhaps of the slightest kind, they apprehend great danger, and even death itself.

Those who are unhappily afflicted by this malady, too frequently become the subjects of raillery and reproof. The strange capriciousness of their complaints induce those around them to suspect them all to be imaginary. This is, however, very far from being the case. It is true, indeed, that from the constant attention they are disposed to pay to every trifling change which arises in their bodies, and from the apprehension with which they are tormented, their account of their feelings may be rather exaggerated, and hyperbolical. But allowing this to be the fact, and even that the most ridiculous and chimerical distresses are imagined by them; even in that case, their sufferings are such, that no considerate person will regard them in any other point of view than as demanding all the solace and relief that friendship, attention, and judgement can bestow (Parkinson, 1801, p. 413).

The author gives a lucid, intelligible account of "Ruptures, or Herniae," and on the subject of trusses, he has this to say:

To those whose circumstances will render the acquisition of this instrument easy, nothing further need be said; but the poor industrious labourer, whose month's earnings would hardly suffice to make the required purchase, should be apprised of the vast importance of the early application of a truss; his better informed neighbour should instruct him, that after the first appearance of this disease, no consideration ought to induce him to make one more laborious exercise, until provided with this security against future mischief: that, if neglected, it will, in consequence of his exertions, increase with rapidity; and may even soon destroy him; and lastly, that there is no little luxury, or convenience, which he ought not to give up; nor any sacrifice of delicacy, which he ought not to make, that prevents his application for benevolent and charitable aid, in procuring this important and necessary instrument (Parkinson, 1801, p. 423–4).

In his remarks on cases of "Rupture at the navel in Children," he states:

. . . the means of cure are in every one's reach; and will be almost always successful, if employed early and attended to continually.

The protruded parts are to be returned, which may be easily done by slight pressure with the fingers; and retained in their proper situation by a conical piece of very soft sponge. . . . This being kept to the part, by the point of one finger, is to be secured by several slips of strongly adhesive plaster, three inches in length, crossing each other in stellated form (Parkinson, 1801, p. 430).

One of the author's favorite topics was the protection of children from accidental injury.

It was the good practice of parents a few years back, to surround the head with a circular quilted pad, covered with silk, which, though not very ornamental, had no forbidding appearance, and must frequently have saved children from considerable injuries to the head. I am so confident of this fact, that I cannot resist the impulse of most earnestly proposing their adoption to the affectionate parent (Parkinson, 1801, p. 443–44).

"The Bite of an Hydrophobic Animal" includes the following paragraph:

From the absorption of the saliva, conveyed by the teeth of the animal into the wound, is the hydrophobia supposed to proceed. The more speedy and the more complete its removal, the less chance is there for absorption. Consequently, in every case of bite from an animal, in which there is reason to suppose hydrophobia exists, the piece should be immediately cut out; so carefully including even the parts immediately surrounding the wound, both at the bottom and sides, that a probability shall not exist of the knife's coming into contact with the wound made by the animal (Parkinson, 1801, pp. 444–45).

"I here conclude the task allotted to me," says the author, writing finis to the first volume, "hoping that, notwithstanding its imperfections, this little work may be of real utility."

The last part of *Medical Admonitions* deals with diseases of children and "Observations on the Excessive Indulgence of Children, particularly intended to show its injurious effect on their health, and the difficulties it occasions in their treatment during sickness."

Under the heading "Epileptic fits," the author tells of an instance of epilepsy he once witnessed in a girl about seven years of age.

The least resistance to the wishes of this little favourite was sure to be followed, by the most violent gusts of passion, which generally terminated in an epileptic fit. After applying for relief to several physicians and employing, without the least appearance of success, every measure which had been recommended; although perhaps, not with so much regularity, and perseverance as the case required, her parents concluded the disease to be insuperable, and declined any further attempts at a cure (Parkinson, 1801, p. 521).

Writing of "Scirrhus and Cancer," the author assures the unfortunate sufferer from a cancer of the breast that there is only one method of cure for this disease: excision at an early stage. But in pre-anaesthetic times, one can conceive with what dread an operation would be contemplated:

It should also be considered that during this state of the disease, the operation is comparatively trivial, and requires but very little time for its performance; that the pain which accompanies it is very far short of that which is imagined; and that the exemption from future disease is rendered is almost certain.

It is worthy of remark, that the opposition to this, and indeed to every operation, does not proceed simply from the dread of the pain; but from an association of the most horrid ideas of every minute circumstance respecting the operation, which can be fancied. The assemblage of the surgeons, the preparation of the instruments, and many other circumstances, the enumeration of which would be opposite to the present purpose, are all represented in colouring so *sombre*, and with touches so masterly and impressive, that the mind is filled with the utmost horror at the picture that imagination has drawn. Reason has no longer power to exert its influence; and the unfortunate sufferer, thus becoming the slave of terror, attempts not to argue and combat with the host of terrific spectres which fancy is constantly placing before the eyes.

When I consider how powerfully the dread of an operation affects the mind, I cannot indeed expect to effect much by argument. Some little good may however, perhaps, follow from this analysis of the ordinary process of the mind in this and similar cases. Since it thus plainly appears, that the dreadful apprehension of the operation is formed, in a great measure, by the mind's dwelling on objects entirely visionary, and that the fear of suffering pain has, perhaps sometimes, the least influence in directing the opposition to the operation recommended. Thus, we often see men who can bear pain with the greatest fortitude, declare, that although they well know the pain of the operation does not exceed that occasioned by the scratch of a pin, yet they can with difficulty bring themselves to consent to suffer the operation of bleeding. Nor can it hardly be doubted, if the most timid female, suffering this disease, was assured, that if she would permit the pain of the part to be augmented exactly to the same degree, and for the same period of time, as it would be by any proposed operation, she should obtain an entire removal of the disease, and even indeed of the part itself, and this merely by an incantation, but that her consent would easily be obtained. But, indeed, it is not necessary to imagine a case, to show that the dread of operations does not depend merely on the fear of pain, as this is sufficiently evident from the much more ready submission to the application of caustics; even under the strongest condition of their producing a much greater degree of pain, and of their affording a considerable less chance of a cure.

The mode of thinking which should therefore be adopted, by those to whom such an operation has become necessary, is to let the mind dwell only on the absolute pain of the operation, abstracted from all foreign circumstances and visionary terrors; and to consider, that a cure may be obtained by only a few minutes increased pain, and this pain very far short indeed of what is most probably anticipated by imagination; but suppose it ever so violent, let it be remembered, that it is but suffering, even say! the most excruciating pain, for a few minutes to procure an entire liberation from misery and despondence.

But unhappily, there are many, whose dread of chirurgical operation is so great, that no argument or consideration can overcome (Parkinson, 1801, pp. 469–72).

The book concludes with an essay on excessive indulgence of children in which the author reveals that he has a sound knowledge of child psychology, and an awareness of the duties of good parenthood. In an "advertisement" he says, "The following observations were originally intended for a periodical publication, *The Monthly Visitor*, a work well calculated to promote the interests of youth. The introductory part therefore appeared in that Work; but, on considering, that a greater chance of their proving useful might be derived from their appearance in a work, more particularly appropriated to medical admonition; they were reserved for the following pages." These observations were later published separately in book form (Parkinson, 1801).

The difficulties to be encountered in examining a child prone to "excessive wailings" is also pointed out, and the author describes how children are terrified by the application of leeches, which he says sometimes results in convulsions. Children, he said, are so tortured by blisters applied to the skin, that they are inclined to tear them off. Children's medicines, he recommends, should always be well flavored. Amongst the ailments peculiar to children he mentions those due to "improper indulgencies in food," and "nervous affections and madness." He specifies those cases in children which require "chirurgical treatment" and makes an impassioned plea that children should all be inoculated against smallpox (Parkinson, 1801).

CHAPTER 6

Medical Literature

James Parkinson's medical writings were mainly articles in the medical press, and tracts, but he did publish another medical monograph besides his "Essays on the Shaking Palsy," viz. *Observations on the Nature and Cure of Gout* (Parkinson, 1805), an octavo volume of 174 pages. In addition to an outline of the pathology and treatment of the disease, it contained a vivid account of the symptomatology culled from his personal experience of gout, as did his father before him (Parkinson, 1805, p. iii).

In the preface, he says:

> Nearly fifteen years ago, I experienced the mortification of finding that I was also under the influence of this tormenting malady.
>
> It was frequently urged, by the hope of benefitting my fellow sufferers, to lay before the public the observations which I had made in this disease, and its several modes of treatment. But, fearing that the work might not prove of that utility which I expected; and knowing that every trifle from my pen must betray the unfavourableness of the moments in which they had been written I was induced to hesitate (Parkinson, 1805, p. vi).
>
> The hope also of rendering an essential benefit to the labouring part of the people . . . was also a powerful incentive to this publication (Parkinson, 1805, p. viii).

His father's case he described as follows:

> Case J.P. was of a sanguine temperament, and born of parents who had not been subject to either the gravel or the gout. His food was generally plain, and his convivial indulgences were by no means frequent, but wine, or some other fermented liquor, was at times generally employed. Until nearly forty years of age he had enjoyed almost uninterrupted health; but at this period he was first attacked by gout in the foot; the fit being uncommonly severe. During the succeeding six or seven years, he was subject to very violent paroxysms, which during the latter three years, occurred twice every year, and confined him at least a month or six weeks each time.
>
> Whilst recovering from a severe and tedious paroxysm he received a visit from the late Dr. Hugh Smith, who informed him that a gentleman of respectability, in Essex, who had been subject to both gravel and gout, had, three years before, taken

Blackie's Lixivium (a solution of pure fixed alkali) for the cure of the gravel, and had never since been attacked by the gout: he therefore recommended to him the trial of it as an anti-arthritic. In consequence of this recommendation, he took the lixivium for a twelvemonth, avoiding wine and other acescents most carefully, no symptom of gout making its appearance. At the end of this time, finding no return of the gout, he relaxed in his use of the soap-lye, and began to drink of stale porter. Within about a twelvemonth, from his thus resuming the use of this powerful acescent, an acute pain took place immediately beneath the pubis, which did not abate of its excruciating violence until two hundred drops of laudanum were injected, mixed with a proper fluid, into the rectum. He continued to endure daily excruciating tortures in this for upwards of two years, notwithstanding very frequent and large doses of opium were employed to abate them. The nature of this complaint was never actually determined, although every circumstance attendant on it seemed, at one time, to warrant the supposition of its being a case of schirrous contracted rectum. At the end of rather more than two years, the pains considerably abated; but very soon after ascites was found to have established itself, the distressing attendants on which closed the melancholy scene (Parkinson, 1805, p. 101–104).

The author then gives a lengthy and detailed description of his own affliction which he was forced to endure whilst writing his geological book, *Organic Remains* (Parkinson, 1804–11) and attending to his professional duties.

J.P. whose father was much afflicted with gout, is evidently of that temperament which is supposed to be most prone to that disease, and always possessed a peculiar idiosyncrasy in consequence of which the taking of any acid, or even acidulous matter, into the stomach, occasioned in a few minutes, a considerable glow of heat in the face and in the extremities. In youth, chilblains, with cramps in the calves of the legs; and through life, coldness, particularly in the evenings, with dryness of the feet, have seemed to point out a languid circulation in the lower extremities. The functions of the viscera forming the *primae viae* have always appeared to be duly performed; no dyspeptic symptoms having ever occurred, excepting very rarely, slight flatulency in the stomach, happening generally for two or three days previous to any arthritic attack. His diet has been generally plain and simple, such as a humble table has afforded: wine has been drank but seldom, and then sparingly, not exceeding two or three glasses at a time (p. 159).

At 38 years of age he experienced the first attack of gout in the ball of the right foot: this, however, was so slight as not to occasion more than a week's inconvenience. In the spring of the second succeeding year, an attack of more violence and of longer duration was suffered in the left hand, the pain and swelling lasting about ten days. In the summer of the year following, having walked a little way out of town, drank nearly a pint of wine, and being exposed to a heavy rain in the evening, he awoke about three o'clock in the morning in such severe pain in the instep of the right foot, as excluded all hopes of regaining sleep. Convinced of its being a gouty affection, and wishing to induce perspiration in the part, he quitted his bed, but found himself entirely incapable of standing on that foot: he, however, drew on a woolen stocking, and regained the bed. The pain now soon became so exceedingly acute as to be almost beyond endurance. Recollecting the case of Mr. Alexander Small, related in the sixth volume of the Medical Observation and Inquiries, in which great relief was obtained by lessening the temperature of the part, he now stripped off the stocking, and laid with it, about ten minutes, on the outside of the bedclothes, exposed to the open air. The swellings on the finger joints now being considerable, and that on the third finger of the left hand manifesting a disposition to inflammation, a leech was applied to it, and the bleeding promoted as much as possible. Almost directly after the bleeding from this joint, the stinging burning sensation began to abate, and was in two or three days removed. In about a fortnight's time the swelling was suspected to be rather diminished; and in about six weeks the diminution was evident: a leech was therefore applied to each of the enlarged joints, and with the same beneficial effects as were produced on the other. The medicinal plan just mentioned being pretty fairly persevered in for about two months, essential benefit was experienced: the joints to which the leeches had been applied were considerably reduced, and even the swellings of the thumb and forefinger, to which the leeches had not been applied, were evidently diminished (Parkinson, 1805).

As to the cause of the gout, Parkinson wrote:

Indulgences in acids is a frequent cause of gout: that an acrimony is then produced, from which is formed a peculiar matter, which is separated and removed from the system, by the inflammatory action of the parts affected with gout; that the process being suddenly checked in the extremities, is liable to be taken up by some other part; whose healthful state, is much more important to the preservation of life, than that which had been primarily affected; that even when no sudden violent affection of any internal part is secondarily occasioned, low chronic inflammation, perhaps of a peculiar kind, seems to be often times produced; and that the ends of the bones, or rather their immediate coverings, become affected in a peculiar manner, in those cases, in which the gouty inflammation has been suddenly checked, or in which circumstances exist which prevent its taking place, notwithstanding the gouty diathesis present (Parkinson, 1805, p. 173–74).

Dr. Robert Kinglake of Taunton, Somerset, published *A Dissertation on Gout* (Kinglake, 1804). His theory was that gout was essentially a local disease, in no way differing from "common inflammation," and not a constitutional disease. Kinglake was the originator of the "refrigeration treatment" of gout, which was in vogue at the time, and which consisted of the application to the affected parts of cloths which had been immersed in a solution of equal parts of cold water and Aqua Ammon. Acet. (Kinglake, 1804).

Parkinson criticized Dr. Kinglake's theory of the causation of gout, and strongly deprecated the cooling treatment which he had tried on himself and his patients. He was not alone in his condemnation of the refrigeration treatment, for he was warmly supported by John Ring, a well known medical practitioner in London, who wrote *An Answer to Dr. Kinglake: showing the danger of his cooling treatment of the Gout* (Ring, 1816). *The Medical and Chirurgical Review* stated that

> Other opponents of Dr. Kinglake's unqualified recommendation of this refrigeration have appeared in the medical journals; and express answers to it from the pens of Mr. Edlin of Uxbridge, and Mr. Parkinson of Hoxton. Mr. Parkinson, therefore, thinks that although Dr. Kinglake has recommended the cooling plan, not doubting its efficacy, or its safety in every case, it is so dangerous as not to be justifiable, either in the inflammatory rheumatism or in the Gout: since ill effects have often arisen from the practice at a distant period of time.

Dr. Kinglake, a doughty controversialist, was goaded by Parkinson's criticisms of his treatment, and replied to them in a book, *Strictures on Mr. Parkinson's Observations on the Nature and Cure of Gout* (Kinglake, 1807a), in which he quoted the favorable opinions of the medical press and fellow practitioners. Another book, entitled *Additional Cases of Gout in farther Proof of the Efficacy of the Cooling Treatment* (Kinglake, 1807b) soon followed.

Parkinson conjectured that the "proximate cause of Gout" was "a peculiar saline acrimony, existing in the blood, in such a proportion, as to irritate and excite to morbid action the minute terminations of the arteries in certain parts of the body." This saline acrimony he supposed to be "the acidifiable base of the uric acid, while the complete formation of this acid depended on some unknown action of some unknown part of the animal system" (Parkinson, 1805, p. 3). Dr. Kinglake wrote "the presence of calcareous concretions (or chalk stones as they are called) are characteristic of ligamentous and tendinous inflammation and not of the influence, as Mr. Parkinson supposes, of *urates* of *soda* in the circulating system."

"Opposition had been conducted with urbanity and candour," Kinglake acknowledged. "It would be a tacit distraction in me," he concluded, "to terminate any strictures on Mr. Parkinson's work, without acknowledging it to be, upon the whole, a medical performance of the first class of respectability" (Kinglake, 1807a).

The first chapter of Parkinson's *Observations on the Nature and Cure of Gout* is devoted to the pathology of the disease, and the second chapter to aetiology and predisposing causes, among which he proposes a constitutional predisposition for the most part hereditary. Among other causes, the author mentions intemperance in the use of spiritous liquors, and intense application to study or business.

Chapter three investigates the relationship of the symptoms to the supposed proximate cause, and chapter four refers to the diagnosis of the disease and its differentiation from rheumatism. Dr. Haygarth, in his description of "Nodosity of Joints" (Haygarth, 1805), thought that this particular form was confined chiefly to persons in the higher and middle ranks of life, but Parkinson thought otherwise and said that he had observed it more frequently among the poor. But, as one reviewer of the book remarked, "The opposition in this case is readily reconciled by the different spheres of practice of the two observers, without impeachment of the accuracy of either."

Chapter six is concerned with the general treatment of cases of gout, while chapter seven recommends the method of treatment to be adopted in an acute attack. Two cases are described, one, "a gentleman of considerable respectability in the City, about forty years of age," who was cured following "a regimen" (abstinence from alcohol, a liberal supply of vegetables, and plenty of exercise) and fifteen grains of sodium bicarbonate a day. The other case, a gentleman about fifty years of age, "who had indulged freely in the pleasures of the table," is described, but his treatment was only partially successful. The author observes that in this case, "the unwillingness to act different from those around him, did not oppose the suggestions of prudence. Hence, as might be expected, but a partial benefit was obtained, the relief however evidently agreed with the degree of strictness with which the rules were attended to."

The final chapter deals with what the author terms "retrocedent gout," and it is here that he condemns Dr. Kinglake's cooling treatment as a potent cause of retrocedent gout, for he says, "the process being suddenly checked in the extremities, is liable to be taken up by some other part; whose healthful state is much more important to the preservation of life than that which had been primarily affected; that even when no sudden violent affection of any internal part is thus secondarily occasioned, low chronic inflammation, perhaps of a peculiar kind, seems to be often produced" (Parkinson, 1805, pp. 86, 88, 89).

Parkinson's medical writings went further than his study of gout. *Hints for the Improvement of Trusses* (Parkinson, 1802a) is the title of a 22-page tract, written in 1802 "for the use of the labouring poor." The author said that he deplored the fact that a truss, a surgical appliance, a necessity for the preservation of life, could be patented and thus become a species of monopoly. In the preface, the author remarks: "So much is it the practice to have recourse to patents, to secure to the proprietor the emoluments of any discovery, of real or pretended utility, in relieving the diseases to which the human frame is subject, that if no apology be necessary, yet some explanation may be proper, why it had been had recourse to in the present instance" (Parkinson, 1802a, p. iii).

He continues:

> The circumstance which renders this explanation almost unavoidable is, that the public has been long in the habit of estimating the value of all such discoveries, by the profits which their proprietors derive from such exclusive patents. Why this practice has not been here adopted will be, perhaps, best shown by stating the writer's sentiments respecting exclusive patents, in any case where the discovery respects the mitigation, or the removal of disease.

Contrivances by which the conveniences, or the luxuries of life are increased, may, perhaps, be considered as fair articles of pecuniary speculation, and of individual exclusive advantage. But ought the necessaries of life to undergo such a species of monopoly? Certainly not. Legislators have seen the baleful consequences, and have, therefore, exacted prohibitory laws, and have subjected those who infringe those laws to severe punishments. But certainly that man is much more deserving of reprobation, who, possessing a knowledge of the means by which a painful, and hitherto fatal, disease may be stopped in its career, unfeelingly beholds it spreading its devastation, far and wide; and instead of anxiously seeking to diffuse the blessed antidote, wherever the disease exists, limits it within the small range of a patent: and exerts himself, only to prevent its beneficial influence from extending to anyone, who may not possess the power of purchasing it of him, at the price which he has affixed to it.

Possessing these sentiments respecting the reservation of exclusive property, in those discoveries which conduce to the preservation of life, and the diminution of disease, it was sufficient to believe it possible, that the present little improvement might, eventually, prove beneficial, to produce such a publication of its description, as might prevent anyone assuming the principle as their discovery, for the purpose of obtaining an exclusive patent. By stating this, however, it is not meant to arrogate the merit of a very important discovery; it is merely offered as a hint, which may probably suggest means of relief, easy of acquisition, in a disease in which, if those or similar means be omitted, a fatal termination may be expected to occur. A hope is also entertained that the principle, capable of being farther extended, may, under the attention of the ingenious mechanic, or even the patient himself, be so modified, as to be applied to the construction of an instrument, still more simple and efficacious than any which has been, as yet, adopted (Parkinson, 1802a, pp. VII-VIII).

The tract describes a rupture, how it is produced, and how it can be recognized. The danger of neglecting a rupture and its consequences are stressed. Of the final outcome of neglect, strangulation with intestinal obstruction, the author says, "it may be proper to remark that even at this period of imminent danger, life may be saved by an operation." He points out that the only hope of cure, or of even checking the progress of the hernia, is by applying pressure by means of a truss. "Among the numerous diseases," he observes, "to which the labourer is exposed, by the employment of those exertions which are necessary to procure him his daily bread, there is hardly one more strongly calls for the aid of benevolence than the malady termed a Rupture" (Parkinson, 1802a, pp. 12,9).

Parkinson's invention is then described:

It is well known that if a piece of stick, placed between the coils of a piece of rope, fastened at each end to some solid body, be turned round a few times, immediately as the force is removed by which it was turned round, it returns to its former situation with considerable force.

A plate which accompanies the description of the construction of the truss illustrates, by means of five figures, different stages in the procedure.

CHAPTER 7

Medical and Surgical Cases

Throughout his life, Parkinson reported interesting, unusual, or dramatic cases to medical societies or journals, the earliest of which are contained in the annual reports of the Royal Humane Society. His father, John Parkinson, was a medical assistant to the society, and the annual report for 1778 states that "Mr. Parkinson, Hoxton, was appointed by the Court of Directors one of the medical assistants for the New River," which signified that he was on call for cases of drowning in that part of the New River which ran through Islington into the reservoir at Sadler's Well, Clerkenwell. After his father's death, James Parkinson succeeded him in this appointment.

Dr. W. Hawes (1730–1808) and Dr. T. Cogan (1736–1818), founders of the Humane Society, and Dr. J.C. Lettsom (1794–1815), were medical assistants whose rota of duty included that portion of the River Thames which flowed between Westminster and London Bridges (Abraham, 1933). The Society paid half-a-crown to the messenger who summoned the doctor, and two guineas to the medical assistant who carried out the resuscitative methods recommended by the Society, provided he continued to apply them for not less than two hours. A further two guineas was paid if the patient recovered. Certain public houses were classed as receiving houses, to which the apparently drowned could be taken, and the landlord was paid one guinea for his services. In Parkinson's area, The King's Arms, Shoreditch, The Blue Coat Boy, Islington, and The Macclesfield Arms, City Road, were receiving houses. The Honorary Silver Medal was awarded to a medical assistant who showed exceptional skill in the saving of life. The Humane Society's Report for the year 1778 states: "Gentlemen presented with the honorary medals for Restoration in the year 1777. Mr. John Parkinson, Junior, for the recovery of the body of Brian Maxey, Oct. 25th, see Case CXLIV. p. 161. Trans. Vol. I."

Parkinson was an apprentice surgeon at the time, and the case was reported to the Society by his father. "Sir," wrote John Parkinson, "It is with the utmost satisfaction I can inform you of a case in which I have been able (by the means recommended by your truly Humane Society) to restore life to one, who, before the institution of your Society, would probably have been numbered with the dead."

Cases of Drowning

About seven o'clock in the evening of Tuesday (Oct. 28th.1777) I was sent for to the assistance of Brian Maxey, aged 29 years (who, as the messenger informed me, had hanged himself). I immediately, with my son, hastened to the place (about a quarter mile distant from my house.)

The poor fellow was to all outward appearances lifeless: the jaw fixed; a coldness had overspread the limbs excepting a very small degree of warmth at the pit of the stomach. A woman, who had some imperfect idea of the means recommended by our society, endeavoured successfully to increase the warmth: and a neighbour had taken from the arm eight ounces of blood. Mr. Dearns arrived and kindly assisted in the inflation of the lungs, friction, &c. for a quarter of an hour; we then perceived an almost imperceptible pulsation: and the pupil, which had hitherto been much dilated, showed a tendency to contract. In forty minutes he fetched a deep sigh, which increased the pulse, and in about an hour the natural breathing took place; and in half an hour more (being an hour and a half after we began our endeavours) his senses returned; he frequently swallowed very diluted brandy: Wednesday 29th. the medicine had operated, but he was exceedingly low, and complained of pain at the pit of the stomach: he had been light-headed during the night, his pulse strong and full. The purging ptisan was repeated, and an erect posture recommended. Thursday 30th. he had three stools; the complaints remain the same; the pulse full and quick. I now thought it necessary to take more blood; the pulse began to sink before three ounces had flowed; a sudorific opiate was prescribed. Friday 31st. the complaints not alleviated in the least; four leeches were applied to his temples. Saturday, November 1st. complaints the same; the sudorific opiate was repeated.

He continued mending until Wednesday, when his chief complaints left him; he is now only afflicted with a numbness of the right side of his head and a dimness of sight.

From all the information I could obtain, I think I may conclude that he must have been hanging half an hour; for he informed me that he committed the action before his wife had been absent five minutes from the room; and from the joint account of his wife and several of the neighbours, more than half an hour elapsed before her return.

Sir, I have only to add, that he expressed the utmost sorrow for his horrid crime, and the greatest gratitude to those who were instrumental in his restoration to his wife and children.

John Parkinson.

Hoxton Square, Nov. 5th. 1777.

In a reprint of the same annual report the letter states that "restoration was carried out by the means you recommend—inflation of the lungs, alternately raising and distressing the chest, for more than half an hour," from which it can be assumed that the original letter had been abbreviated, and that it also contained a reference to young James Parkinson's active participation in performing artificial respiration. Doubtless the father and Mr. Dearns had both written reports which must have influenced the Directors of the Society in awarding the Honorary Silver Medal to the young apprentice, who at this time was 22 years of age. Two years later, James himself reported a case of drowning to the Society, which is thus recorded in the annual report for 1779, "Case CCXI. No. 20. Copy of a letter from Mr. Parkinson, to James Horsfall, Esq; dated Hoxton Square, Sept. 7th. 1779."

Sir,

On Monday evening last, the bearer of this (Timothy White) happily saved a boy (John Scott) fourteen years old, from drowning: the boy was bathing in a pond near Bethnall Green, and finding himself beyond his depth, cried to his companions for assistance; they being able to afford him none, ran to this man, whom they saw at two fields distance, who whilst he was running towards the pond, saw the boy rise and sink again three times; the last time he remained under the water three or four minutes; he now dived and luckily brought him up, but entirely insensible, and so remained for five minutes. If you think the man deserving of the award adjudged for these cases, I doubt not to its being gratefully received.

I am, Sir,
Your most obedient humble servant,
James Parkinson.

P.S. It may be necessary to inform you, that I was the boy as soon as brought home; he then being universally affected with convulsive twitchings, his pulse being extremely irregular, intermitting (and that for a long interval) every four or five pulsations: he likewise complaining of a violent pain at the pit of his stomach with urgings to vomit, I gave him a gentle emetic, and after it a sudorific anodyne; the next morning he was well, excepting some little pain at the pit of his stomach.

Parkinson wrote several letters of recommendation, one of which was as follows:

Sir,

I am happy to inform you of the restoration of Anne Dinah, a child about fourteen months old, who last Monday fell with its face down into a tub of water, where it lay near ten minutes, and when discovered showed no apparent signs of life, its face being livid and much swelled. Sarah Hosell, the bearer of this, and the person who discovered the child, having obtained some idea of the mode in which such cases should be treated, rubbed the child with salt until, after a while, signs of life began to appear, when the child was wrapped in blankets; at this time I first saw the child, when its breathing was very laborious and interrupted, her face still much bloated, and the pulsation of the artery at the wrist hardly perceptible. I then used such methods as appeared to me most proper, to call back and fix the still wavering and vital flame—happily, with success.

Sir, I am yours,
With the utmost respect,
James Parkinson.

The child's parents live in York Street, Webb Square, Shoreditch: its father is a weaver. The mother was going to lay it out, and would have done so, if Elizabeth Quince had not interfered and insisted on rubbing it with salt.

Effects of Lightning

Parkinson was a Fellow of the Medical Society of London, founded by Lettsom in 1773, and his first paper was read before that society on February 4, 1789. It was printed in the *Memoirs* of the Society (Parkinson, 1789) under the title "Some Account of the Effects of Lightning."

The paper gives an account of a storm in which two men were struck by lightning at the same time in Crabtree Row, Shoreditch, one being in his house and the other walking on the street outside.

The first man, who had been outside, had been taken for dead. He had recovered his vital functions gradually, but these remained impaired. His head was thrown back and he was unable to move it. The lower part of his body was without sense or movement. He coughed blood and complained of pain in the chest and head. His extremities were blackened. His right side was marked with a large red streak with forklike ramifications.

Parkinson treated the man with bleeding, hot brandy and water, and wet flannels, after which he slept soundly and woke much recovered.

The second man was less seriously injured. His extremities were also blackened, and he had several red, forking streaks as well as small burns. He received the same treatment as his fellow victim, and responded as positively.

The house, when examined, showed signs of damage such as broken tiles, melted irons, and broken window frames. A peculiar circumstance relating to the blackening of the victims' limbs was observed: when rubbed, they turned a deathly white but, after the rubbing stopped, the discoloration returned. The red streaks, when rubbed, became lighter in color, and stayed that way.

Trismus

On June 18, 1811 Parkinson read a paper before the Medical and Chirurgical Society, of which he was also a member, entitled "A Case of Trismus" (Parkinson and Parkinson, 1811), which was actually written by his son

John William Keys Parkinson, M.R.C.S. John Harkness, a surgeon of Broad Street, Ratcliffe, had previously reported to the Society "A Case of Trismus following a contused wound of the head," and in a letter which appeared in the Transactions, he wrote: "A case of a similar kind was treated some time since, in nearly a similar way, by Sir William Blizard (1743–1835) and Mr. Parkinson, Junior of Hoxton, Mr. Parkinson has been so good as to draw it up, at my request, for the Society."
The case is described by John Parkinson, as follows:

Mrs. D., a lady of 50 years of age, of a spare habit and delicate constitution, received on 19th Sept., a compound fracture of the leg, for which she was attended by Sir William Blizard and myself. Three weeks after the accident, when the wound had a perfectly healthy appearance, and the bone had acquired a considerable degree of firmness, she was suddenly attacked with pain and stiffness about the back of the neck, a difficulty of swallowing, and an inability to open the mouth to a greater extent than about half an inch. Her pulse was small and weak, and her skin dry, but not hot.

There was no dyspnoea nor abdominal hardness; and a stool, which had been procured the day before, did not present anything particular in its appearance.

A drachm of tincture of opium was given soon after the accession of these symptoms, which in a short time produced a diminution of the pain in the neck and of the difficulty in swallowing, and a slight degree of relaxation of the muscles of the jaw. She was ordered to take five grains of calomel, and to continue a drachm of tincture of opium every hour.

By eleven o'clock the next morning, she had taken two ounces of tincture of opium without producing any sensible effect, except that of keeping her free from pain in the neck, and producing a slight degree of relaxation of the jaw. It was therefore ordered to be repeated in the same dose, and at the same periods.

The tincture of opium was found absolutely necessary to keep off a general uneasiness and restlessness, which had hitherto never failed to come on, when it had been occasionally omitted for five or six hours.

At the expiration of three weeks, the tincture of opium was gradually diminished; and, after a few days, it was found necessary to give it only at night. The bowels now began to resume their normal action, and only required the occasional use of a cathartic.

In five weeks here health seemed to be completely re-established. The only inconvenience she experienced, was from the continuance of some degree of stiffness of the muscles of the jaw, which still occasioned a slight difficulty in opening the mouth, and which did not entirely go off for many weeks afterward (Parkinson and Parkinson, 1811).

Appendicitis

On January 21, 1812, James Parkinson read a second paper before the Medical and Chirurgical Society, which was also written by his son John, entitled "A Case of Diseased Vermiform Appendix," and which is of particular interest since it is now confirmed that this was the earliest reference to appendicitis in English medical literature.

"A preparation of diseased vermiform appendix in my possession," wrote J.W.K. Parkinson, "was removed from a boy about five years of age, who died under the following circumstances."

He had been observed for some time to decline in health, but made no particular complaint until two days before his death, when he was suddenly seized with vomiting, and great prostration of strength. The abdomen became very tumid and painful upon being pressed: his countenance pale and sunken, and his pulse barely perceptible. Death preceded by extreme restlessness and delirium, took place within twenty four hours.

Upon examination, the whole surface of the peritoneum was found inflamed, and covered with a thin coat of coagulated lymph: and slight adhesion had taken place between the peritoneum covering the viscera, and the parieties of the abdomen. The viscera, independent of the inflammation of their peritoneal covering, appeared in a perfectly health state, excepting the appendix vermiformis of the caecum; but about an inch of its extremity was considerably enlarged and thickened, its internal surface ulcerated, and an opening from the ulceration, which would have admitted a crow quill was found at the commencement of the diseased part, about the middle of the appendix, through which it appeared that a thin dark coloured and highly fetid fluid had escaped into the cavity of the abdomen.

Upon opening the appendix, a piece of hardened faeces was found impacted in that part of it which lay between the opening, and that portion of the appendix which was not evidently marked by the disease (Parkinson and Parkinson, 1816).

Hydrophobia

In February 1814, Parkinson sent the following letter, addressed "To the Editors of the London Medical Repository," the leading medical periodical of the period, on the subject "Cases of Hydrophobia." Parkinson detailed two cases, the first of which was as follows:

Gentlemen,

I transmit you such particulars as I am able respecting the cases of Hydrophobia, from which must have proceeded those vague reports lately in circulation. Having parted with the notes of the first case in expectation of its publication and being no longer in possession of those of the second, I shall be obliged to rely, in the following account, entirely on my memory, which is however so impressed with the most important facts, that although it has to refer to a rather distant period, it will not, I trust, materially mislead me.

It was to the best of my recollection, about the year 1780, that I was called in the afternoon to a servant maid of the Rev. Mr. Clare, Queens Road, Hoxton, of about 28 years of age, affected with soreness of the throat and inability of swallowing. On visiting her, she informed me, that the evening before she had first felt some slight soreness of the throat, but that now she had no power of swallowing. Whilst relating this, her countenance and manner manifested such peculiar marks of distress, as led me, although I had never seen the disease, to suspect its existence in the present case. These suspicions were unhappily soon confirmed. On presenting to her a glass of water, she watched its approach with much anxiety, and immediately on its touching her lips, she shuddered extremely, became strongly convulsed, and appeared to be affected with the highest degree of horror and alarm.

The advice of Sir William Blizard was immediately requested; and the time, nearly an hour, which elapsed before his arrival, was employed in watching the symptoms, and obtaining information respecting the origin of the malady.

The pulse was frequent; the heat but little more than natural; the mouth was dry, and the tonsils slightly inflamed. She complained exceedingly of thirst, but said it was impossible to drink; for that, as I had seen, directly as the liquid touched her lips, it threw her into convulsions; I therefore promised that I would give her some drink without letting it touch her lips. This I attempted by dropping water on her tongue from a teaspoon, she opening her lips and mouth as wide as she was able, and leaning her head back on the pillow. This succeeded for a little time, she swallowing the water, and expressing much gratification; but a drop falling aside on her lips, she became exceedingly agitated, and was with difficulty persuaded to allow me to make another attempt, in a mode in which I told her I could better secure the water from dropping aside. This, however, she at length submitted to, and by dripping the water from the end of my finger on her tongue, she swallowed nearly half a glassful of water with much pleasure, and without any apparent inconvenience. I then placed some calves feet jelly on her tongue, in its more solid form; which, as it melted, she swallowed pleasantly, and of this she took freely.

An opportunity offering, I inquired what had become of the little dog I had been accustomed to see at the house. She informed me "that it had become so troublesome that it had been sent away." "Had it", I asked, "become cross and snappish?"—"No", she said, "never—it was rather too fond, for it was always perpetually leaping on her lap, and licking her hands and arms" which I observed were still rough from chaps. "Then", I said, "he never bit any of the family?" "Never." "Pray", I then asked, "were you ever bitten by any dog?" "Never", she eagerly answered, adding, "Why do you ask?".

When Sir William came, he procured her immediate removal to the London Hospital, where to the best of my recollection, a consultation of all the medical and surgical officers was obtained, but in vain. She died, I think within forty-eight hours from her removal. The inspection of the body, which was performed with a particular degree of minuteness and care, presented only the usual unsatisfactory appearances.

Every inquiry respecting the dog proved unsatisfactory; nor do I believe that any reason appeared for supposing that the dog had manifested any signs of sickness before, or after, its removal. In consequence, I considered this as a Case in which the disease had occurred spontaneously, or that it had been communicated by an animal in a state, which, according to the knowledge we then possessed, was supposed to be incompatible with the existence of the disease.

Under these circumstances I with much pleasure delivered up my notes to Dr. Marshall, in consequence of his expressing a wish immediately to publish the case. The publication was, however, I believe, never made (Parkinson, 1814).

Only a year later an account of the case was published in a book by Andrew Marshall entitled *The Morbid Anatomy of the Brain* (Marshall, 1815).

The author, a well-known teacher of anatomy in London, stated that the material for the publication was collected from the papers of Dr. Marshall (who died in 1813) by S. Sawrey, who was his assistant.

Part II of the book gives "A history of two cases of canine madness, with observations on the seat and nature of the disease." "The one," says the author, "was the case of a young lad, brought into St. Bartholomew's Hospital; the other patient was a young woman, received into the London Hospital." "The earlier state of the young woman," the author continues, "after she was taken ill, was observed by Mr. Parkinson, surgeon, in Hoxton, who, besides, made some accurate notes on the nature of her malady, as will be mentioned in the proper place" (Marshall, 1815, p. 19).

There was an interval of 26 years between the occurrence of this case and the account of it which Parkinson wrote in *The London Medical Repository*. It is not surprising, therefore, to find some discrepancies between the report which he made at the time and sent to Dr. Marshall and the article which he wrote from his recollection of the case after the lapse of so many years.

Fevers

Parkinson was much perturbed by epidemics of typhus fever in Hoxton. In 1813, a particularly bad outbreak of typhus had swept through the courts and alleys which led out of the Kingsland Road. Owing to overcrowding of the houses, many in the same house were afflicted, and the populace was so frightened that hired nurses would not attend to the sick, and neighbors shunned them. In consequence, he had persuaded the Guardians of the Poor to erect a fever block in the workhouse infirmary, the building of which was carried out in 1819. Nearly five years later he witnessed another epidemic of the same disease, but on this occasion he had all the facilities of the fever wards at the infirmary at his disposal, and in an article published in the *London Medical Repository*, entitled "On the Treatment of the Infectious or Typhoid Fever," he demonstrates the better results obtained by improvements in treatment and hospital care and nursing (Parkinson and Parkinson, 1824). John W.K. Parkinson collaborated with his father in the writing of this article:

The fever here referred to commences with a rigor or shivering, followed by loss of muscular power, increase of heat, pains in the head, loins and limbs, the tongue, which at first was white, becoming loaded with a dark brown fur. The gums also become foul, from the adhesion of a tough mucus, which extends even partly over the teeth. As the malady advances, the heat becomes considerably augmented; petechiae, and vibices frequently occur; and, at length, the powers of the nervous system failing, every faculty becomes impaired, and death, preceded by convulsions, ensues.

This variety of reported cases demonstrates the range of Parkinson's interests and practice. Only the observations on appendicitis are still remembered, but many of his concerns, such as resuscitation techniques or the free admission of fresh air for fever victims, show him to have been a man who kept up with the latest medical thinking of his time.

CHAPTER 8

The Parish Doctor

As a result of his association with the unpopular reform societies, Parkinson suffered greatly from social and professional ostracism. At the outbreak of the war with France in 1793, he had indicated that he was an uncompromising pacifist, and matters were made worse when the French Revolution reached its climax with the outbreak of the Reign of Terror.

His zeal for reform continued, and he began to take an interest in the affairs of his own parish, being elected a parish councillor, or Trustee of the Vestry for the Liberty of Hoxton in 1799, as the following entry in the Minutes of the Vestry indicates:

> To elect and appoint fit persons, inhabitants of this Parish, to be trustees in the place of those who have died, removed out of this Parish or refused to act. The following person was proposed and having a majority in shew of hands was declared duly elected (viz.) Mr. James Parkinson in the room of Mr. Thomas Stiff deceased.

The minutes for the year 1825 denotes that, "In Hoxton Liberty, Mr. James Parkinson, deceased. Mr. James Currie in the room of Parkinson" so that for 25 years he was a member of the Vestry, which was responsible for the upkeep of the highways, paving of sidewalks, illumination of streets, maintenance of paupers, and the support of the parish church of St. Leonard. Parkinson is mentioned as having been chosen as a Collector of the Poor Rate for the year 1799.

As a body representing all parishioners, the Vestry was not concerned only with local affairs but on occasion exhibited concern for matters of national importance. Since few had a parliamentary vote, the views of the people could be manifested only by petitions to Parliament. Town meetings, open to all, were sometimes held to protest about grievances affecting the nation, e.g., the inhabitants of Shoreditch were loud in their protestations against the harsh and unchivalrous treatment of Queen Caroline by her husband, George IV, when a bill was brought before the House of Commons "to deprive Her Majesty Queen Caroline Amelia, of the title, prerogatives, privileges, and pretensions of Queen Consort of the Realm, and to dissolve the Marriage between His Majesty and the said

Queen." George IV, as Prince of Wales, had openly flouted his mistresses before the Court so that the people were well aware of his extravagances and failings.

When the war with France entered a critical stage in 1798, the Government was hard pressed for additional soldiers and sailors. The militia regulations now imposed on the Vestry the onus of finding substitute militiamen, a task almost impossible to accomplish, not least because the soldiers' wives and families had to be kept by the parish out of the poor rates. As in the case of Queen Caroline, indignation was expressed by a petition conveyed to the House of Commons by their Member of Parliament.

Despite the failure of the reform societies (to several of which Parkinson belonged) to effect changes in the House of Commons, Parkinson was still a strong supporter, 25 years after first joining the London Corresponding Society. His zeal never diminished for the cause he was convinced was right.

At a meeting of parishioners on February 13, 1817, it was resolved:

> to take into consideration and determine upon the most prompt and proper means to be adopted and pursued (in conjunction with the Cities of London and Westminster) for obtaining a redress of the numerous grievances under which the country groans and a full real and true representation in Parliament of such portions of the English people as are either misrepresented or not represented.

It was also resolved:

> that the following Gentlemen be the Committee and that they are now authorised to carry into effect the above resolutions in such a manner as may appear to them most expedient. Messrs. Sharpe, Flower, Jacobson, Lawrence and Parkinson.

The "above resolutions," which were unanimous, read as follows:

> That the Civil Governments were instituted avowedly for the benefit and advantage of the people and should naturally secure the greatest possible share of individual liberty at the least possible expense. But the Laws of England, once the pride and admiration of the Civilised World—now fettered by legal fictions and prohibited by taxation have been gradually undermined from the time of the pensioned Parliament of Charles II to those of George III, until at length entirely subverted, scarcely a trace of liberty remains.
>
> That the Constitution which is said to govern our unhappy country is formed of the Irish admixture of three principal forms of Government, Monarchy, Aristocracy, and Democracy, so accurately poised that Justice Blackstone has laid down, "When the Independence of any one of the three should be lost or that it should become subservient to the views of the other two, there would be an end to our Constitution" and Montague has written "That whenever the Legislation shall become more corrupt than the Executive, England will lose her liberties by an Act of Parliament." These opinions were fatally verified and our unhappy Country has a long time borne the Yoke of a Tyrannic Oligarchy.

> That in this degenerated and subverted state of our Constitution lies the true cause of our present misery and distress and that nothing short of our original Constitution can restore the Liberty and Prosperity of our Country.
>
> That our only object in petitioning for a reform in Parliament is to secure a real representation of the people according to the original intention of our Constitution, and there to restore that admirable and necessary Counterpoise to the other two branches of our Government.
>
> That we most solemnly deprecate the practice of rejecting the petitions of the people merely because the language is not suited to the Ears of those members whose Corrupt Practice we hold in the greatest abhorrence.

Sir Francis Burdett, M.P., had tried to revive the question of parliamentary reform by presenting a bill in 1809 before the House of Commons asking for a change in representation, but this was defeated by a large majority, only 15 members supporting the bill. Burdett was imprisoned in the Tower of London for publishing a pamphlet in which he described the members of the House of Commons as "a part of our fellow subjects collected together by means which it is not necessary to describe." Henry Brougham, M.P. for Middlesex, in whose constituency Shoreditch stood, was a supporter of Burdett's bill, so the petition of the people of Shoreditch was presented by a member of the House who sympathized with the case for reform. One can detect the hand of Parkinson, still an ardent advocate of parliamentary reform, in formulating the petition since the arguments were precisely those he had put forward in his political pamphlets published over 20 years previously.

He was elected to the Board of Trustees of the poor of the parish of St. Leonard's and, in the minutes of January 28, 1803, it is stated:

> Mr. Parkinson gave notice of his intention of bringing forward a motion at the next meeting that the state of the apprenticed poor be taken into consideration.

In particular, his great concern was the welfare of some 70 children who had been bound as apprentices to masters or mistresses by the Trustees in neighbouring parishes, lest they should become chargeable as paupers within their own parish on reaching adult life. This was a vicious circle because neighboring parishes adopted the same methods and apprenticed their children in Shoreditch.

Having been moved to act because a child apprenticed by an adjacent parish had been murdered by her Mistress, Parkinson proposed a resolution on March 11, 1803, calling on the Board of Trustees to investigate the conditions under which the children lived and worked. He told the Board that it was a sad reflection that no law existed whereby the duties of a Master or Mistress could be defined, and no inspectors had ever been appointed whose duty it was to visit apprentice children to report upon their condition. The general behavior of employers to their charges also needed careful investigation. He persuaded the Board to nominate members who would voluntarily make periodic visits to the homes of "the defenseless apprenticed poor at the mercies of their Masters and Mistresses."

When the committee gave its report on July 19, 1803, the very first visit of inspection showed how Parkinson's allegations of ill treatment by the employers were fully justified:

> In performing the task they have undertaken, this Committee are under the painful necessity of relating one shocking instance of seduction and depravity, in doing this, they trust they shall clearly show to this Board the necessity of its continued attention to those children who generally know no other guardians.
>
> Jane Lang whose fate has led the Committee to make this suggestion was apprenticed in the year 1799 to John Conway of No. 119 Petticoat Lane, Whitechapel, and was seduced by her said Master, who, the Committee are informed, was a married man and the father of six children—she proving with child, the Master ran away—and the poor creature was delivered of a child in the workhouse of St. Mary Whitechapel, and died, about six months since.
>
> To diminish the uneasiness which reflection on so melancholy a circumstance must excite, the Committee have deferred to this place the particular account of some of their inquiry, the conduct of the Masters and Mistresses of whom claim their most honourable mention.

The Committee divided the apprentices into three categories: the first contained the names of those particularly well treated, the second the names of children who could not be traced, and the third comprised the names of those who "need the interference of proper officers of the Parish and by their occasional attention the state of the children may be materially amended."

The Board, its conscience stirred, decided to act quickly and at their next meeting resolved "that a Committee be appointed to consider the best means of securing the comfort of the Parish Apprenticed children." In consequence of the Committee's deliberations, a regular system of visitation of the apprentices was assured, and they were inspected twice yearly.

Parkinson was on the rota of visitors for the year 1806, and his report of visits made, alone, to the houses of 72 of the parish apprentices showed how time consuming a task it was for a busy medical practitioner in the days of slow transport.

This system of regular and searching enquiry produced the desired effect and, six years later, the minutes noted that "Mr. Parkinson reported—that the Officers, himself and several of the Committee, visited the children apprenticed and found them in general comfortably situated." Parkinson drew up a list of regulations for the apprenticing of children, which was as follows:

> With the hope of preventing the continuance of such evils, seven regulations are proposed (See Appendix I).

It must have been gratifying to Parkinson to have witnessed the satisfactory end of his unceasing labors for the welfare of the apprenticed pauper children, orphaned, friendless, or deserted. He also witnessed the passing of Acts of Parliament which ameliorated their unhappy lot and which incorporated many of the changes he had himself originated in Shoreditch.

At a meeting of the Trustees of the Poor held on July 1, 1813 (Parkinson was not present), "a motion was made and seconded that Messrs. James Parkinson and Son be elected Surgeon, Apothecary and Man-Midwife to the Poor of the Parish during the pleasure of this Board, which on being put, was carried unanimously—and—they were declared duly elected accordingly." The resignation of James Parkinson as Trustee was read and it was resolved the same be accepted. James Parkinson, and his son John William Keys, who, like his father, was a Member of the Royal College of Surgeons, thus became the parish doctors.

One of Parkinson's duties as Parish Doctor was to act as House Surgeon (the appellation then given to the workhouse medical attendant) to the parish workhouse, and another to attend to and give medical relief to sick paupers in their own homes. After but a year as parish doctors, Messrs. Parkinson and son complained of the inadequacy of their salary, and on November 3, 1814, the minutes state "That the salary of Messrs. Parkinson and Son was not adequate to their services" and at the next meeting on December 1, 1814, it was resolved: unanimously "That the Salary of Messrs. Parkinson and Son be increased from £156 per annum to 250 guineas per annum from Michaelmas last."

In October 1815 the Board received a letter from Messrs. Parkinson and son expressing their anxiety concerning an epidemic of typhus fever which had raged in Shoreditch in the previous year (see Appendix II).

The recommendation for separate fever wards was a progressive move, as at that time the only fever wards in a general hospital were those established by Dr. Haygarth at the Chester Infirmary in 1783. The only fever hospitals in England were the Houses of Recovery in Manchester and Liverpool, and the London Fever Institution in Islington.

The Board, ever mindful of the public purse, reduced the sum recommended, i.e. £650, to "a sum not exceeding £450." Though it took nearly four years of argument and persuasion to get the Trustees to agree to Parkinson's scheme, it took only four months to complete the erection of the building, which comprised two fever wards, one for males (six beds), and the other for females and children (12 beds).

This was another triumph for Parkinson, as these were the first separate fever wards to be set up either in a general hospital or a workhouse infirmary in the Metropolis of London. This innovation in Shoreditch was soon followed by the opening of fever wards in the workhouses of the parishes of St. Pancras and St. Giles.

In the year 1818, a Select Committee of the House of Commons on Contagious Fever listened to evidence given by physicians from the teaching hospitals of London which showed how advanced Parkinson's conceptions were.

Dr. E. Roberts, physician to St. Bartholomew's Hospital, said that it was not good practice to segregate fever cases and that there was no fever ward at his hospital. Furthermore, he thought that typhus fever was not communicable. Dr. Thomas Young of St. George's Hospital stated that his hospital too had no fever ward, and he also was not in favor of segregating fever cases. "I should think it probable," he observed, "that the disease being more concentrated, it would more easily be communicated to the nurses and physicians." Dr. H.H. Southey of the Middlesex Hospital stated that there was no fever ward in his hospital either, and he thought the danger of mixing the fever cases with other sick was trifling. Dr. Yelloly of the London Hospital said that his hospital also had no fever ward, but acknowledged that many nurses, porters, visitors to patients, and one doctor had died from typhus fever contracted in the hospital. The physician from the Westminster Hospital also said that his hospital had no fever ward. Dr. Alexander Marcet of Guy's Hospital was the only doctor in favor of them and hoped that Guy's Hospital should have them.

In this year, 1818, an epidemic of typhus fever swept through the workhouse of St. Mary's, Whitechapel; inmates (both adult and children) and attendant were equally affected. The mortality rate was so alarming that Drs. John Yelloly and Isaac Buxton of the London Hospital were called in to investigate and report on the epidemic.

Whilst the Trustees were still considering the matter of the fever wards, Parkinson published a pamphlet entitled "Observations on the Necessity for Parochial Fever Wards," which was published in 1818. Unfortunately, no copy appears to exist today, but we can gather a notion of the contents from reviews of the publication (see Appendix III).

The equipment of the fever wards was brought up to date. The plain wooden planks covered with straw were replaced by "16 new wrought iron bedsteads with sacking and headboards for the Fever Wards at the Workhouse, at a cost of £27.4.0." The heating of the wards was modernised by the purchase of "three elliptic stoves," at a cost of £9.5.6.

At a meeting of the Board held on November 9, 1819, "the following Rules and Regulations for the Fever Wards at the Workhouse, recommended by Messrs. Parkinson and Son" were approved (see Appendix IV).

These rules can be regarded as a serious and praiseworthy attempt to combat the spread of infectious disease by isolation and segregation of infected cases. The rule prohibiting visitors is significant, as at that time there was no restriction of any kind put upon visitors to hospitals, workhouses, or even jails.

The minutes of a meeting of the Board held on Thursday, December 23, 1825, records the death of James Parkinson in a letter from his son:

Gentlemen,

With feelings of deep regret I have to announce to you the Death of my much beloved and respected father.

He was suddenly attacked on Sunday afternoon with a severe paralytic affection which deprived him of his speech and the use of his right side. Every means were employed for his recovery but they proved of no avail—he expired on Tuesday morning at half past ten o'clock.

I have the honour to be, Gentlemen,
Your most obedient servant,
J.W.K. Parkinson

James Parkinson died at 69 in harness, the Parish Doctor to the date of his death, and it was whilst acting in this capacity that he wrote his classic "An Essay on the Shaking Palsy." Gout, which had tormented him for so many years, was followed by a degeneration of his arteries which hastened his end. In his remaining years, he had moved from No. 1 Hoxton Square to No. 3 Pleasant Row, Kingsland Road, which is where he died on December 21, 1824.

Since the partnership now no longer existed, his son, J.W.K. Parkinson, continued to act as Parish Doctor alone until the year 1829. On July 21, 1833 he sent a letter of resignation to the Board, in which he said:

In resigning the situation of Medical Attendant of the Poor of this parish, I beg leave to offer you my grateful acknowledgements for the confidence you have for so many years reposed in me and to assure you that I should not desert a post of which I was always proud (whilst I could conscientiously discharge its duties) had not the number of applications for medical assistance so much increased of late as to ensure the more efficient attendance on the sick poor.

Following his resignation, no less than six medical men were appointed as parish doctors to carry on the work formerly done by two. Not long afterward, he moved to 11 Suffolk Place, Lower Road, Islington, a distance of about three quarters of a mile from Hoxton Square, where he died in 1836. It can be inferred from the contents of his will that he practiced as a surgeon at this address.

Appendix I: Regulations for Apprentices

1. That no child should be apprenticed before the age of 12 years unless to a relation, or under such circumstances as a Board of Trustees shall think render it proper.
2. That no child shall be apprenticed to Cotton Mills or any other employment by which there does not appear to be a probability of such child gaining a livelihood which in time is expired.

3. That children shall be apprenticed in the Parish whenever proper masters can be obtained.

4. That notice shall be given by posting Bills through the Parish of the resolution of the Trustees, that the parishioners shall be preferred as masters for the children.

5. That masters shall convenant that the apprentices shall not work on Sundays—that they shall be sent to some place of public worship, once at least, every Sunday—that they shall not oblige them to work longer than twelve hours in any one day, and not before six in the morning, nor after nine o'clock in the evening—that they shall furnish them with cloathing on the 1st day of May in every year;—that they shall produce the apprentices thus cloathed before a Board of Trustees on the 1st day of May in every year—that they shall not oblige more than two to sleep in one bed—that they shall give notice to the vestry clerk of their removal from their place of dwelling—and that as the Master and Apprentices together cannot vacate the Indentures, the Master shall give immediate notice to the Vestry Clerk of the apprentices quitting his service.

6. That the apprentice fee be increased to Five Pounds, two of which is to be paid at the time of Binding, and one pound on the 1st day of May for the first three years, when the child shall be brought before the Board of Trustees.

7. That every child on being apprenticed have a printed paper given to him or her containing an account of the Regulations under which he or she is bound with the name or address of the vestry clerk to whom he or she may apply in case of any supposed Injury from the master.

Resolved that the foregoing regulations which have been read be adopted.

Appendix II: Letter for Fever Wards

Gentlemen,
We take the liberty respectfully to state that from the present occurrence of Putrid Fever in this Parish, the adoption of measures for the security of the inhabitants, and of the public in general by preventing the spread of contagion appears necessary.

In a few months of the last year, more than a hundred cases occurred in the Kingsland Road. In one of the courts hardly a room escaped the visitation; and in some rooms five or six persons suffered its attacks. Recourse was had to the Fever Institution; but from its distance, and the number of patients which were at first attacked, its aid proved insufficient to the due Separation. The mortality was however less than could have been expected, and from the attention of the overseers in obtaining the cleansing of some of the courts, and from their liberality in affording every assistance to the sick, whilst the necessary measures were employed for the destruction of contagion the disease declined.

For some months past very few cases have occurred and of these several have derived assistance from the Fever Institution, and at present we do not know that it exists in the district. But from the miserable state of a great part of the dwellings of the poor, and from the frequent change of their tenants, by which infectious diseases are perpetually like to be introduced, it is to be apprehended daily that the disease may again return.

We therefore most earnestly recommend that the Trustees take these circumstances into their most early consideration and resolve on the adoption of some measure by which the poor attacked by Fever may be immediately separated from the healthy and in a detached Building—a *Fever Ward* under such regulations as may produce an entire exclusion of visitors, and secure a strict separation of the sick from the healthy and an annihilation of contagion.

Should the Gentlemen Trustees be disposed to attend to the recommendation, we will take the liberty to remark that by the adoption of a warm bath, great benefit might be yielded in many cases of disease, besides its utility for the purpose of cleanliness especially in many paupers at their first admission into the house would be very considerable.

Jas. Parkinson
J.W.K. Parkinson

To the Trustees of the Poor of St. Leonard, Shoreditch, Oct. 12th 1815.

Appendix III: Reviews of Pamphlet on Fever Wards

The European Magazine stated that:

Mr. Parkinson very judiciously recommends the revival of what have been called pest-houses, for fever as well as small pox. He also adduces some very important hints for the reflection of Magistrates (pp. 18,19) viz., the spread of contagion, arising from the commitment of vagrants to crowded gaols, without directing the necessary precaution of isolating them from other prisoners. The matter of this pamphlet is highly good; not false philanthropy.

A review of the pamphlet appeared also in the *London Medical and Physical Journal* [1819 no. 241 (3 of vol. XL1)], under the heading "Medical and Philosophical Intelligence" which is a valuable summary of the contents:

We shall devote a portion of the present Number of our Journal to a subject solely of local interest, which has been brought before the public by Mr. James Parkinson; that of the propriety of establishing fever wards in the different parishes of the metropolis. . . .

. . . Mr. Parkinson, who appears to have given much attention to the subject . . . thinks that the disease under consideration may be referred to a fever that occurred in the inhabitants of some houses in a court in Kingsland Road, in the parish of St. Leonard, Shoreditch; whence he traces it to its present epidemic state, . . . the measure urged by Mr. Parkinson with so much laudable zeal, is sufficiently obvious. The numerous scenes of wretchedness that he has witnessed in consequence of the ravages of the prevalent fever, where the victims have suffered every privation, in consequence of their inability to procure hired nurses, and the fears of their neighbours depriving them of gratuitous domestic aid, are such as cannot fail to excite the attention of all those who may become acquainted with them, and lead to exertions for the prevention of similar calamaties in future.

'The first measure which should be adopted,' says Mr. Parkinson, 'is that of establishment of fever wards. It is, indeed, astonishing in an age distinguished for its flow of charity and kindly consideration for the poor, that the warning voice of the Society for bettering the Condition of the Poor, and the anxious remonstrances of numerous respectable and benevolent individuals, have yet produced the establishment of only one place, furnished with accommodation for about seventy patients, for the reception of the pauper fever patients of London and its environs.

An establishment with eighteen beds (six for males and twelve for females) would, it is expected, be sufficient for the paupers belonging to a parish whose population is from 40,000 to 50,000 persons; and the beds and the building might always be usefully employed as an infirmary, when the fever was subdued.'

After having demonstrated the evil, and indeed, the fatal consequences, of either keeping fever patients in their lodging houses, or of sending them to the poor house, Mr. Parkinson briefly points out the advantages of placing them in a fever ward.

'Immediately as a patient is affected with fever,' he observes 'he is removed thither by a proper conveyance, his tatters immersed in water, and he is carefully washed and confined to this ward: in which would be secured the means of ventilation and purification, the patient would gain every chance of his recovery, whilst the air of the ward itself would be preserved innoxious, and consequently the attendants would be exposed to much less danger. The patient, on his recovery, would be supplied with fresh clothing, and would return to society, no longer conveying about with him the poison of pestilence.'

'Satisfied,' continues the author, 'that those to whom these lines are more particularly addressed are anxious to promote, on every occasion, the comfort of the poor and the welfare of society, not the slightest fear is entertained of all due attention being paid to these suggestions, offered with combined earnestness and respect.' After pointing out the several advantages of minor importance that would ensue from the adoption of the measure he advises, Mr. Parkinson concludes with remarking that, 'should every parish provide fever wards for their own poor, the necessity would still continue for the liberal support of the more general institution of this kind. Where else could be sent the servant, or other inmate, who ought not to have parochial relief claimed for them; but who, for the safety of the families with whom they reside require immediate removal and separation from the healthy? Where could be sent the wretched and unprotected Irish Poor, suffering here, in their own country, as it is called, all the privations and misery of exile?'

Lond. Med. and Phys. Jour.
Vol. XL1 1819
p. 269-271

Appendix IV: Rules for Fever Ward

1. Patients on being brought in to have their skin well washed with warm water and soap, and if strong enough to be put in the warm bath, whilst washed. If the hair is in considerable quantity it is to be reduced.

2. Their clothing on being taken off to be immersed in cold water, rinced, hung to dry, and then preserved for the patient, on going out, in separate, marked parcels.

3. The patient, after being cleansed, to have fresh linen, and put to bed each singly, when possible.

4. The bedsteads, recommended to be of iron, the beds of straw, the pillows of waste-rush, and the rugs of cotton.

5. The patients to be kept out of their beds as much as circumstances will allow.

6. The attendants to be a Male and Female Nurse, the male for the men's, and the female for the women and Children's wards—these attendants to receive and clean the patients, and to take charge of their cloathing, and to deliver it to them, on their going out.

7. As the taking of spirits must augment the injurious effect of the Infection, and therefore improper both for the patients and the nurses; no spirits are to be admitted into the Fever Wards, unless under such particular circumstances as may, in the opinion of the Surgeon, warrant their employment.

8. That no visitor be allowed, but by the permission of the Surgeon; not even of any person already in the house.

9. The carriage for the sick to be always kept ready, and two men (the same persons, to be always employed) to be appointed as porters for the same, allowing them 6d. for each turn.

10. On the removal of a patient, the porters to leave a printed request that the apartment from which the patient has been removed, may be duly cleansed, and white-washed.

    FORM FOR THE PRINTED NOTICE

    A person affected with Fever having been removed from hence to the Fever Ward of St. Leonard, Shoreditch, the Parish Officers most particularly request of the proprietor of the apartment, that it be well cleansed, aired, and white-washed.

    (signed) ______________________ House Clerk.

11. The foregoing rules to be regarded as part of the general regulations of the Workhouse; and any breach of them, by either patients or nurses, to be punishable according to the appropriate clause in the local Act of Parliament.

CHAPTER 9

James Parkinson's *Chemical Pocket Book*

At the end of the 18th century there were several good textbooks of chemistry available to English readers, though most of them were translations of French works. The principal ones were the English translations of books by A.L. Lavoisier, (1743–1794), A.F. Fourcroy, (1755–1809), J.A. Chaptal, 1765–1832),[1] and William Nicholson's *First Principles of Chemistry* (Nicholson, 1796). However, these were all substantial works and Parkinson, believing there was a need for a short introduction to the subject, published *The Chemical Pocket Book* (Parkinson, 1800a) in 1800.[2]

In his preface, Parkinson described the book as an "assemblage of chemical facts"; which he hoped would be particularly useful to beginners in the subject. He certainly presented a great deal of information in his 244 small pages of text,[3] dealing with the elements and their compounds in a fairly conventional order: oxygen, hydrogen, and nitrogen being followed by the earths (lime, magnesia, etc., not yet known to be metallic oxides), the two solid alkalis (soda and potash, also not known to contain oxygen), the principal acids (believed to be oxides of nonmetals such as sulphur, carbon, and phosphorus), and then the 21 known metals. The information was up to date, with many references not only to the books of Lavoisier,

---

[1]A.L. Lavoisier, *Elements of Chemistry*, translated by Robert Kerr, 1st English edition, Edinburgh, 1790; 4th edition, Edinburgh, 1799; A.F. Fourcroy, *Elements of Chemistry and Natural History*, translated from the 4th French edition by R. Heron, 4 vols., Edinburgh, 1796; J.A.C. Chaptal, *Elements of Chemistry*, translated by W. Nicholson, 3 vols., London, 1795.

[2]According to G.S. Boulger in the *Dictionary of National Biography*, vol. 15 (1904), p. 315, the book was published in 1799, but this must be incorrect for Parkinson included (pp. 213–16) an account of Humphry Davy's latest work, taken from the February 1800 issue of Nicholson's "Journal." The incorrect date, 1799, is repeated by P.A. Gerstner in the *Dictionary of Scientific Biography*, vol. 10 (1974), p. 323.

[3]The last page is numbered "216" but there are in fact 244 pages of text; apparently due to a printer's error, the page numbers 152–169 are used twice, the real p. 169 being followed by pp. 152*–169* and then by p. 170.

Chaptal, and others, but also to recent papers in Nicholson's *Journal of Natural Philosophy, Chemistry, and the Arts*, the *Philosophical Transactions* of the Royal Society, and various foreign periodicals. His preface included acknowledgements of his indebtedness not only to the authors whom he cited, but also to George Pearson (1751–1828) and William Babington (1756–1833), who lectured in chemistry at St. George's and Guy's hospitals, respectively. Parkinson was apparently acquainted with them and may have attended their courses.

The first edition of the *Chemical Pocket Book* contained 216 pages of a size known as 12 mo., and subsequent editions which appeared in 1801 and 1803 were considerably enlarged. The fourth, and last, edition (1807) was of octavo size and contained some 412 pages. The frontispiece of this edition bears an engraving by the artist Springsguth, the upper half representing a few pieces of chemical apparatus, captioned, "The Economical Laboratory of Guyton," whilst the lower half shows columns of chemical symbols, designated "The Chemical Characters of Hassenfratz and Adet." A full explanation of the illustrations is given in the book. On the title page is depicted an upturned glass vessel within which is a brightly burning flame, and beneath it the words "iron wire burning in oxygen," for a description of which the reader is referred to page 143. This edition contained "the latest discoveries" and included "A Sketch of Mr. Davy's recent discoveries of the Chemical Agencies of Electricity, and of the decomposition of the fixed alkalies of Ammonia, the Alkaline Earths etc."

In the preface to this edition the author observes:

> Although it is thus hoped, that this little work will convey a general and correct idea of the science, it is not, however, expected to supersede the use of some other elementary works, differing, in some respects, in their object, or in their mode of conveying the proposed instructions.
>
> The obligations of the Editor to the best Chemical writers of the age are obvious: it is, however, necessary to particularize that these memoranda have been enriched by a careful collation with the course of lectures on Chemistry, delivered at the Royal Institution of Great Britain, by Mr. Davy. Whilst thus pointing out those to whom his acknowledgment of obligations are due, gratitude and honest pride impel him, respectfully, to mention the names of Wollaston, Chenevix, Hatchet, Babington, Crichton and Powell (Parkinson, 1800a [4th ed., 1807], p. V–VI).

Parkinson had stated, in a letter written to Dr. Richard Watson (1737–1816), Bishop of Llandaff, that his interest in chemistry began with the reading of *Chemical Essays* (Watson, 1781–87), which the Bishop wrote when he was Professor of Chemistry at Cambridge. After the publication of the first edition of the *Chemical Pocket Book* the author probably listened to the lectures which Sir Humphry Davy delivered at the Royal Institution of Great Britain and which were inaugurated at the very beginning of the

19th century. It thus appears that he had studied chemistry earnestly and assiduously for a period of 20 years before settling down to write a textbook of the science, and by this time chemistry had been considered an essential subject for a medical education. In his book *The Hospital Pupil* (Parkinson, 1800), he modestly recommends his own book on chemistry as an introductory textbook for medical students.

However, there is no record of his having written an article on a chemical subject. The only contribution of an original nature to the science was his discovery of the muriatic acid test which demonstrated the presence of animal membrane in marble fossils. The action of muriatic acid on marble was a classical experiment illustrating the first of Fourcroy's Laws on Chemical Affinity: "Chemical Affinity takes place only between bodies of a different nature." Parkinson had accidentally discovered that sometimes the organic matter of a fossilized animal had not been completely replaced by inorganic substances in the process of fossilization.

His little book he acknowledged as simply an "assemblage of chemical facts" collated from numerous sources, as the multitude of references recorded in the book seems to indicate. It was essentially a theoretical textbook containing only a few details of experiments, but it was a complete summary of what was then known of the science, presented in an interesting manner. After a perusal of contemporary textbooks on the subject, the *Chemical Pocket Book* appears to be more expressive, more intelligible, and more easily readable than the others. He brought the subject to the level of the ordinary reader, and its popularity was probably due to the simplification of the science for the benefit of the beginner. Definitions of chemistry given by Dr. Joseph Black (1728–1799), Dr. T. Thomson (1733–1852), and Fourcroy for example, were not free from objection, being somewhat ambiguous and illogical, and whilst these definitions are so confusing, Parkinson's own definition is much more lucid, though brief and less comprehensive. "Chemistry," he says, "is the science which discovers the constituent principles of bodies, the results of their various combinations, and the laws by which those combinations are effected" (Parkinson, 1800a, p. iii, 1).

Like most chemical writers, Parkinson followed Lavoisier in describing heat as an elementary substance named "caloric" but he also stated, without comment, the contrary opinion expressed by Humphry Davy (1778–1829), Thomas Beddoes (1760–1808), and Benjamin Thompson (Count Rumford) (1753–1814), that it was a phenomenon caused by the motion of minute particles of matter. He adopted the chemical nomenclature that had been proposed by Lavoisier and other French chemists in 1787, using Pearson's English translation of 1794, revised in 1799. Pearson originally translated the words "gaz" and "oxyde" and "gaz" and "oxyd," but in 1799 he altered these to "gas" and "oxide," as well as making other changes.[4] Parkinson's early chapters were printed before he learnt about these changes, and he adopted them in the latter parts of the book, which included contemporary references to topics discussed earlier. This must have confused some readers, although the reasons were fully stated in the preface.

The sections of the book which follow are headed, "Of Alkalies," "Of Sulphur," "Of Neutral Salts," and "Of Carbon." Concerning the latter, he says that "Carbonic Acid Gas, formerly termed *fixed air*, or *aerial acid*, was discovered by Dr. Black, it is formed by the combination of carbon, oxygen, and caloric." Dr. Black called it fixed air, and Bergman later showed it to be an acid which he termed aerial acid (Parkinson, 1800a [2nd ed. 1801] p. 46).

"Of Muriatic Acid," "Of Acid of Boras," and "Of Fluoric Acid" contain nothing of particular interest. "Of Metallic Substances," however, contains the names of 21 metals then known to chemists, namely, platina, gold, silver, quicksilver, copper, iron, lead, tin, zinc, antimony, bismuth, cobalt, nickel, manganese, uranite, sylvanite, titanite, chrome, arsenic, molybdenite, and tungstenite. "The three first," he says, "are called *perfect*, or *nobel* metals, and the others *imperfect*, or base." Under iron, he remarks that "Mr. Mushet, on the suggestion of the Editor, observes that carbon exists in steel, in a concrete state, though not crude—in chemical union, however, and not in mere mixture as in crude iron (*Phil.Mag.*)." Did David Mushet (1772–1847), ironfounder and metallurgist, who effected so many improvements in the manufacture of iron and steel, carry out some experimental investigation at Parkinson's instigation? (Parkinson, 1800a [2nd ed., 1801], p. 67, 87.)

"Of Bituminous Substances," "Of Stones," "Of Volcanic Productions," "Of Vegetable Substances;" as a geologist the author writes on subjects with which he is completely familiar. With respect to stones he describes the composition of calcareous stones, limestone, marble, chalk, tufa, calcareous spar, stalactites, stalagmites, flosferri, pisolites, hamites, roestone, and swinestones, whilst under "Volcanic Productions" he gives a masterly description "Of the Primeval State of the Earth, and its subsequent changes."

"Of Animal Substances" concludes the book. "The constituent parts of animal bodies," he says, "consist of but a small number of radical principles; their various natures chiefly resulting from the different proportion in which these are combined. These radicals appear to be *nitrogen*, *carbon*, *hydrogen*, *phosphorus*, *oxygen*, and *lime*." Some 38 substances are described as to their chemical composition. Among them secretions of the body such

[4]For an account of Pearson's and Parkinson's nomenclature and symbolism, see M.P. Crosland, *Historical Studies in the Language of Chemistry*, 2nd ed., New York, 1978, pp. 195–96, 254.

as the gastric and pancreatic juices, urine, bile, sweat, tears, saliva, and blood; structures of the body such as the bones, cartilage, hair; food stuffs such as milk and eggs; and the prussic, zoonic, formic, and bombic acids, which are obtained from animals. Very little, however, was known about these substances at the time so most of them have very brief descriptions (Parkinson, 1800a [2nd ed., 1801,], p. 192–93).

The study of the structure and composition of the blood had not proceeded very far by the year 1801.

> The Blood, supports life, by supplying all organs with the peculiar juices they demand.
>
> The blood, when at rest, coagulates; and then separates into a yellow liquid called *serum*, and a clot or *coagulum* . . . .
>
> The whole of the blood, which by anatomists is divided into serum, red globules, and coagulating lymph, is found, when chemically examined, to consist of *albumen*, *gelatine*, and *fibrine* . . . fibrine may be regarded as that part of the blood which has undergone the most complete animalization; and from which the muscular fibre and other organs of the body are formed.
>
> The blood is found to derive its colour from iron, combined with phosphoric acid, reduced to a state of a sub-phosphate of iron by the soda of the blood.
>
> It seems that the colour is produced by . . . i.e. the oxidizement of the iron, during the passage of the blood through the lungs. The blood which has just circulated through every part of the animal, and has been brought back by the veins to the heart, is propelled into the lungs of a dark red colour, and impregnated with hydrogen and carbon. By inspiration the lungs are distended with air, the oxygen of which combines with the carbon forming carbonic acid; and with the hydrogen, forming water; another part of the oxygen unites with the blood, which returns from the lungs, and passes into the arteries of a bright red (Parkinson, 1800a [4th ed., 1807], pp. 297–299).

The chemical composition of urine occupies several pages, and much of the information had been compiled from William Nicholson's translation of A.F. Fourcroy's textbook, *A General System of Chemical Knowledge* (Fourcroy, 1804), which was considered to be the most comprehensive and up-to-date treatise on the subject. Parkinson defines *urine* as, "an excrementitious fluid, secreted by the kidneys, in its natural state it is transparent, of a peculiar smell, a citron yellow colour, and a saline taste." He described the urine which is voided after copious drinking as aqueous, having hardly colour or smell, and called it *crude urine* or *urine potus*, whereas that which is voided after a full meal possesses all the characters of normal urine and is called *faeces sanguinis*. Urine, if allowed to stand, undergoes spontaneous decomposition, loses its characteristic odour, and acquires the smell of ammonia (Parkinson, 1800a, p. 185).

Some alterations in the composition of the urine in certain diseases are described as follows:

> In *gouty* habits the urine contains much less phosphoric acid than is in healthy urine; but it becomes increased during the paroxysm. In *general dropsy* it is loaded

with albumen: but in *dropsy from diseased liver* no albumen is present. . . . In diabetes Mellitus Dr. *Rollo* has proved the urine to contain a very large proportion of saccharine matter, depending on an hyperoxygenized state of these system resulting from a morbid state of the stomach, and peculiar combinations formed in it (Parkinson, 1800a [4th ed., 1807], p. 311).
Urinary calculip were supposed by Scheele to be chiefly composed of what he termed *lithic acid*. . . . Dr. Pearson termed it the uric oxide. (Other calculi had been found to be composed of urate of ammonia, phosphate of lime, ammonium-magnesium-phosphate, and oxalate of lime.) (Parkinson, 1800a [4th ed., 1807], p. 312.)
Dr. Pearson has found the uric acid in arthritic concretions. Dr. Wollaston discovered that they were composed of that acid and soda. Fourcroy and Vaquelin confirmed this analysis, finding them to consist of Urate of soda with a considerable quantity of animal gelatinous matter.
Concretions, such as are found in the pineal gland, salivary glands, and as are discharged from the lungs, are in general formed of phosphate of lime (Parkinson, 1800a [4th ed., 1807], p. 314).

No mention is made of any chemical tests which would show the presence of abnormal substances in the urine (sugar, for example), though there is this observation concerning sugar in the blood which appears to show that a test for sugar existed: "Dr. Wollaston has ascertained the non-existence of sugar in the blood of diabetic patients."

Obviously *The Chemical Pocket Book* sold well, for it was reprinted in Dublin in 1801 and a second edition was published in London in the same year. Its 250 pages of text, only 6 pages more than the first edition, were generally unaltered but included references to more recent work. This edition was again a success and was reprinted in Philadelphia in 1802 with an appendix by James Woodhouse (1770–1809), Professor of Chemistry at the University of Pennsylvania. A third edition of 272 pages was required in 1803. Parkinson now enlarged his account of minerals and described the use of the principal reagents for the analysis of mineral waters. He also devoted several pages to a completely new topic, the chemical effects of electricity, taking his account mainly from that published in 1802 by J. Bostock, but including some information from the printed syllabus of the course of lectures that Humphry Davy gave at the Royal Institution in 1802. Davy gave this course, always brought up to date, every year, and it is unlikely that Parkinson failed to attend it at least once, though the list published annually by the Royal Institution shows that he never became a member of subscriber.

The earlier editions of the *Pocket Book* contained tables of chemical affinity, drawn up by Pearson, who based them on earlier tables by T.O. Bergman (1735–1784), which showed the order in which substances displaced each other in chemical reactions. These were retained in the third edition, but Parkinson now added other tables, showing, for example, the compositions of common compounds and the temperatures of various freezing mixtures, which must have added to the utility of the book.

Beginners in chemistry, seeking factual information, evidently still turned to Parkinson's book, for a fourth edition appeared in 1807, but by then it was facing formidable competition from other elementary books. The first of these was *An Epitome of Chemistry* (Henry, 1801) by the Manchester chemist William Henry (1774–1836), first published as a small volume of 216 pages—about the same size as Parkinson's—in 1801, reaching a fourth edition in 1806 and a fifth in 1808 before being transformed in 1810 into the more substantial *Elements of Experimental Chemistry* (Henry, 1810) which was frequently revised until 1829. In the preface to the early editions Henry commented on Parkinson's *Pocket Book*, stating that "a very cursory examination . . . will evince that the plan and objects of the two books are totally different." This must have been obvious to any well-informed reader, for the *Epitome of Chemistry* is clearly written by a man with considerable personal experience of chemistry who, unlike Parkinson, did not merely assemble facts from other writers.

Two other popular books appeared in 1806, both written in the form of a conversation, which must have proved attractive to beginners. One was *A Chemical Catechism* (Parkes, 1806) by Samuel Parkes (1761–1825), a manufacturing chemist, which ran to 12 editions during the author's lifetime. The other was *Conversations on Chemistry* (Marcet, 1806) by Jane Marcet (1769–1858), who published no fewer than 18 editions before her death.

It may have been the presence of these competing books which caused Parkinson to delay the fourth edition of his book until 1807. Now a substantial volume of 368 pages, it was still rather inaccurately called *The Chemical Pocket Book*. His list of acknowledgments now included the name of Richard Powell (1767–1835), physician to St. Bartholomew's Hospital from 1801 to 1824 and founder of the Literary and Philosophical Society there; Parkinson was presumably a member of his circle. Much new material was included in this edition and it had, he said, been carefully collated with Davy's course at the Royal Institution. Still, he juxtaposed contradictory statements from different authors without comment, and this must have been noticed by readers. For example, he again gave a Neptunist account of the formation of minerals from an aqueous medium, but immediately followed this with a summary of the Plutonist theory that they originated from a molten mass, citing the works of James Hutton (1726–1797) and Sir James Hall (1761–1832). He expressed no preference for either theory.

Humphry Davy's early electrochemical researches had been described in his first Bakerian Lecture, read to the Royal Society on November 20, 1806, and Parkinson was able to give a short account of it in an unnumbered leaf printed after the main text and bound in at the end of the 1807 edition of the *Pocket Book*. The book was still in print in 1809, unlike the early editions which had sold out quickly, and Parkinson then reissued it with the new date "1809" on the engraved title page and a much longer account, filling pages 369–412, of "Mr. Davy's recent discoveries." This was, however, the last appearance of the *Pocket Book*[5].

This little book on chemistry met a long-felt need for a concise yet comprehensive work, inexpensive yet adequate for the requirements of the beginner, which could be easily and rapidly read and re-read with the object of committing it to memory. It was, so to speak, the science of chemistry in a nutshell. Parkinson's book appears to have been superseded by that of Samuel Parkes, which reached some 13 editions, and Mrs. Marcet's *Conversations on Chemistry.*

---

[5]There is a copy of the 1807 issue of the 4th edition in the library of the Wellcome Institute for the History of Medicine, London; the 1809 issue is in the British Library.

CHAPTER 10

The Madhouse Doctor

James Parkinson had a wide and varied contact with lunacy. For at least 30 years he had been the visiting doctor of a madhouse at Hoxton, kept by John Burrows, and after his death by his widow, Mrs. Ester Burrows, and her son George William. This private asylum consisted of two sections, one for those who could afford the full cost of maintenance and the other for pauper patients sent by the Guardians of the Poor and supported by the Poor Rates of their own parish.[1]

For over 200 years, Hoxton had achieved an unwanted notoriety from its private madhouses. Satirical writers of the 17th century made Hoxton an object of their wit, and the clowns at Bartholomew Fair could always raise a laugh at the mention of the name. Andrew Marvell, referring to Parker, Bishop of Oxford, wrote that "he is fit for nothing but Bedlam or Hogsdon."

In the late 17th and early 18th centuries, nearly all the lunatics in the metropolis of London and its environs were accommodated in the three private madhouses situation in Hoxton. As late as 1819, of a total of 1551 certified lunatics in London and its seven-mile radius, 1351 were accommodated in madhouses at Hoxton and Bethnal Green. Pauper lunatics are excluded from these figures since they were, at this period, not subject to certification and registration with the Commissioners in lunacy.

In those days lunatics were sometimes chained, to benches by day and to wooden cribs at night. Handcuffs on wrists and manacles on ankles with connecting chains were in common use, while whips were at hand to beat the defenceless wretches into submission.

Maniacs and melancholics, the incontinent and those of dirty habits were herded together, despite biannual inspections by the Fellows of the Royal College of Physicians appointed Commissioners in Lunacy, and the anxious concern of visiting relatives and friends. The keepers were sometimes hardened drunkards and ex-gaolbirds, brutal ruffians with sadistic propensities.

---

[1]The Poor Rates were taxes, levied on property holders, used to support the poor.

Hoxton House, unlike Whitmore House, received pauper patients as well as private. From the year 1792, officers and men of the Royal Navy who became insane were confined here. Inmates of Newgate and other prisons who became mad were transferred there until they either recovered their sanity or died. Inspectors of Naval Hospitals inspected the asylum at intervals and a naval surgeon visited and treated the Navy men. Pauper patients were seen by the various parish doctors and private patients by their own family doctors, but there was also a house surgeon and apothecary who visited daily.

In the first quarter of the 19th century, the public conscience was awakened and it was realized that, although conditions were bad in the private asylums of the metropolis, they were not much better in the two public charitable madhouses, Bethlem in Moorfields and St. Luke's Hospital in Old Street. The Inspectors of Naval Hospitals had called attention to the state of the naval lunatics, and a Select Committee of the House of Commons summoned a large number of witnesses in 1815 to give evidence concerning the treatment of the Navy men in Hoxton House. The Guardians of the Poor in certain parishes, which included at least two members of parliament, and certain benevolent philanthropists, had long been dissatisfied with the treatment of the pauper lunatics. At last a book, written by a former visiting doctor employed by Thomas Warburton, exposed the shocking conditions, and the Government reacted by setting up a Select Committee in 1827 which thoroughly investigated the harsh treatment of patients. Deaths from neglect and maltreatment had occurred but, because of bribery, these cases were rarely reported to the coroner and even the coroner's juries were corrupted.

The three Hoxton madhouses stood within a few hundred yards of each other. Holly House, better known as Burrows', was undoubtedly the best managed of these and the patients were there treated with a more humane consideration for their welfare. Less is known about this asylum, in contrast with all the other houses in Hoxton, for the simple reason that neither Mrs. Burrows, the proprietor and superintendent, nor James Parkinson, the regular medical attendant, were ever summoned to give evidence at a Select Committee, and the members of the Committee sometimes referred to Burrows' House for favorable comparison. Holly House consisted of three houses adjacent to each other. It would have taken Parkinson about five minutes to walk from his residence in Hoxton Square to Holly House.

Edward Wakefield (1774–1854), a social reformer who devoted his life to the welfare of lunatics and consequently visited a large number of madhouses throughout the country, described a visit paid to Mrs. Burrows' house in these words:

> The next place I went to see was kept by Mrs. Burrows, a widow, and her son, at Hoxton; this is an old house extremely inconvenient for the purpose, but still the parish paupers are kept distinct from patients who pay at a different rate, and

there are several attempts at classification. . . . I examined some of the men in a convalescent state, as to their food, who stated in the strongest way that they had plenty . . . there were the same sort of trough beds where the patients were chained at night.

Sir James Williams and Serjeant Pell, Guardians of the Poor of the Parish of St. Pancras, following an interim visit, reported that in comparison with other madhouses, they found "Burrows at Hoxton cleaner, better managed, less crowded."

The Commissioners in lunacy reported the impressions made at their visits as follows: "Nov. 17th, 1816 – greatly too much overcrowded." "Nov. 16th, 1821 – the houses are old, and ill-accommodated to their present purpose, and the pauper divisions are too much overcrowded." The Commissioners also added that the Houses lacked proper warmth and comfort due to the scanty fees paid by the parishes. Although the average bed occupancy was 118, at one time there were 140 patients in the establishment, of which 100 were paupers.

Three interesting cases concerning patients who were under Parkinson's care in Holly House have been recorded, and all must have occasioned his much trouble and anxiety. The first concerns a young man who lived in Shoreditch, named George Davis. When Parkinson gave evidence to the Privy Council on the subject of the Pop-Gun Plot, he wrote:

Immediately on quitting the Council Chamber, I went to the Horse Guards, on the request of an almost distracted mother, to make enquiries regarding her son, a lunatic patient of mine, who had been enlisted in a newly raised regiment in that state. From thence I went to Hemel Hempstead, in Hertfordshire, where the regiment then was, but failed in obtaining his discharge, although the medical gentleman of the place, as well as myself, certified his being a *violent* and *dangerous* lunatic. In that shocking state, bound in a strait waistcoat, he was therefore conveyed to Liverpool. I mention these circumstances, because it occasioned my absence from town for two or three days (Smith, 1795).

Fuller details of the case can be gathered from newspaper reports. The *Courier*, under the heading "Crimping," made the following observations:

We are sorry to observe that the most censurable conduct is still discoverable amongst the persons who are employed in enlisting men for His Majesty's service. A transaction which has just taken place proves the truth of this remark.

A young man, George Davis, of Shoreditch, who had been repeatedly out of his mind, had been removed out of a proper place of confinement, through the false tenderness of his friends, before he had recovered his senses.

Immediately on his liberation, he escaped from his attendants, and within less than a week, enlisted in a new regiment (the Birmingham Blues). His mother being informed of this step, and that he had been shockingly treated and otherwise ill used by the recruiting party, entreated the assistance of a medical gentleman, Mr. Parkinson, who resides in the neighbourhood, where he had been confined, and who had attended him whilst in a state of confinement.

Accompanied by a proper keeper from the madhouse, they went on Wednesday last to Hemel Hempstead, in Hertfordshire, where they found the poor creature in a lay-loft, his face much bruised, one eye almost closed from a blow, and to add to the mother's distress, she was informed that, under the idea that he was *shamming Abraham*, he had experienced the horrid discipline of flogging with a whip, and beating with a stick.

A medical gentleman, Mr. Cherrington, who lived in the town and declared him a lunatic as soon as he saw him a day or two after his being enlisted, had kindly interfered the evening before, put him in a strait-waistcoat, and told the Quartermaster at his peril to remove him.

He now joined Mr. Parkinson in a certificate declaring him to be a dangerous lunatic, and that the worse consequences might result from his not being put in a proper place of confinement; the mother, in the utmost agony, intreating he might be put in the care of the keeper, and conveyed to the house whence she had taken him.

But tears, remonstrances, and certificates were useless; the poor wretch was taken in that stage of raging madness, torn from his distracted mother, placed in an open cart between two soldiers and carried to join the party at St. Albans.

At his arrival at St. Albans, he appeared to be so much injured by his journey, that his distressed mother feared he would die upon being removed from the cart, she therefore renewed her prayers, but was informed that all she could say would be of no avail, as he would be sent in the evening to Birmingham (*Courier & Evening Gazette*, 1794).

Nearly two weeks later, the *Courier* published a letter to the Editor from the young soldier's mother:

Sir, I am exceedingly sorry your kindly stating the cruel treatment my unfortunate son has experienced, shall subject you to the imputation of admitting into your paper a *falsehood*, and that for such a purpose as to *exasperate the people against the military*.

The account which appeared in your paper of Monday, the 13th, was true in every particular, except that the gentleman who sent it, gave a more favorable statement than might have been drawn up: since all that was intended was to induce some benevolent character, who possessed influence, to furnish the proper mode of recovering an unfortunate son, who with a raging distempered brain, was dragged from a mother who would have expended her last farthing to obtain his security, and exposed to the miseries of a common soldier.

The following circumstances being added to the former account, you will be able to judge of the sufferings of one whose mind is in such a condition, as to make her fear she may soon be in the same state with her poor distracted son. On my arrival at Hemel Hempsted, I found that independent of his state of insanity, there was great reason to suppose my son to have been irregularly inlisted: that he had been transferred out of one regiment to another: that the marks of ill-treatment on his face, and the accounts I received of his ill-usage were such as to warrant the supposition that his sudden removal was to prevent any examination of the injuries he had suffered.

On his arrival at St. Albans, when I found my poor son, who no longer knew his distressed mother, was to be dragged away I knew not whither, I could not restrain my tears and exclamations, which, however were soon drowned by the singing and hootings of the soldiers.

Should any gentleman who possesses the power, be induced, by this statement, to step forward and furnish an unhappy mother with the means of ascertaining whether her son is in existence; or, if possible, if he be yet alive, to give her the opportunity of securing him from insult and ill-treatment by putting him in a proper place of confinement, they will confer an obligation never to be forgotten, and may be furnished with particulars respecting this melancholy circumstance, by condescending to apply to

Margaret Davis

No. 13 New Inn-yard, Shoreditch
(*Courier & Evening Gazette*, 1794).

The son was not heard of again and it is presumed that he had died. It is not surprising that Parkinson loathed the methods of the crimps and press gangs who forcibly kidnapped young men and forced them into a service they disliked.

A second interesting case of insanity is described in Parkinson's own words. "A gentleman farmer was brought to a house for the reception of lunatics, his friends grounding the necessity for his confinement on his conducting his affairs in such a manner as must soon bring him to ruin. On speaking to the patient, he said, if his friends could state any circumstance which he could not defend on principles of reason and equity he would consent to be confined for the rest of his days. He was then asked:

Do you not give more wages than other farmers? Yes. Why do You? Because I am of opinion that then standing wages of labourers is much too small; and the neighbouring farmers agree with me in that opinion, but have not integrity enough to follow my example, although they know their labourers to be almost starving.

But have you not had it clearly demonstrated to you, that this preceding must terminate in your ruin? Yes, but—a question in my turn. Am I to be deemed a madman because I will not save myself from ruin by starving a number of my fellow creatures? Well, but your friends say, that you have thoughts of leaving your farm to your servants, and to make a tour over Scotland, setting out with only a crown in your pocket. Is that a rational intention? Yes, I have certainly a right to make what tour I please; it will be a more rational tour than your sparks of quality make, for I go to inform myself of the agriculture of the country I pass through.

But you leave your farm at the mercy of your servants? So do other farms, and more madly than I should, since, by generosity I have assured myself of the fidelity of my servants. But was it not madness to think of setting out on this excursion with only a crown in your pocket? So, extravagant generosity is first brought as proof of my madness, and, this failing, you mean to prove it by my parsimony.

But I can explain this part of my conduct also. I know I injure myself by the wages I pay, and therefore can spare but little for myself—so much for my parsimony. But how is this crown to carry you through? Thus—I shall take one of my horses for the first thirty miles, and then travel on foot for the next twenty; and thus, with care, my five shillings will carry me fifty miles from home. Now the object of my journey is agricultural knowledge, and my wish is to obtain it as cheap as I can; therefore

> I will hire myself as a labourer until I have got five shillings more, and then set off again. I have got such recommendations as will ensure my employ and extra wages. In this manner, I shall perform my tour; and get, perhaps, as much useful knowledge as will enable me to pay my men without incurring ruin.
>
> Staggered by the acuteness of these answers, the medical gentleman was with difficulty induced to sign the certificate of his lunacy, and, at last, did it with that want of strong conviction which left it a burden on his mind.
>
> In a little time all doubts were removed; he threw himself over the balustrades of a staircase, although with but little injury. On being asked what induced him to do this, he said, he long had it in intention, and had only waited for God's consent: that he, that morning had put a piece of paper on the frame of the window to ascertain whether his intention was approved. If the paper blew outwards, he was to infer he had permission; and if inwards not.
>
> Well, he was asked, did it blow inwards? No, he answered, it remained where I placed it, from which I concluded the answer was I might do as I liked, and therefore I threw myself downstairs. Let it be considered, that if, in the first instance, the medical gentleman had refused to certify him as a madman, and any serious mischief had followed, the heaviest reproaches would have been heaped upon him; and a disinterested opinion, delivered according to his judgment might have seriously hurt his professional character (Parkinson, 1811, pp. 6–8).

Parkinson's professional reputation was assailed in a third case, that of Mary Daintree, who was confined in Holly House by a certificate of insanity signed by him. Considering the ease with which the sane, and the harmless insane, were detained in the private madhouses, actions at law for assault and false detention as lunatics were uncommon in those days. The mercenary motives of the proprietors, the eagerness of some grasping medical men to certify on the slightest pretext of lunacy, and the evil disposition of relatives who wished to rid themselves of encumbrances, laid them all open to prosecution for illegal detention. A certificate of insanity had to be signed by either a physician, a surgeon, or an apothecary, but the status of the latter had never been defined by law, so that any quack could call himself an apothecary and sign such a certificate without question.

Law reports of the period contain many cases in which the keepers of Hoxton madhouses were sued for illegal detention of lunatics. The wily Warburton was prosecuted several times, but by a combination of cunning and perjury, was always fortunate enough to escape punishment, emerging victorious from such actions by pleading justification and posing as a public benefactor who had shielded the public from the menace of lunatics who would otherwise be at large. Judges and juries accepted this defense, despite disclosures that Warburton himself had signed the certificate though not a medical man, even when he had detained a patient without a medical certificate of any kind and merely upon the entreaties of relatives or friends. In the celebrated case of Chawner v. Warburton, where the latter was proceeded against for "trespass, assault and false imprisonment," the plaintiff, the Rev. George Chawner, had been seized, admitted into Whitmore House, and therefore detained for seven years without a medical certificate of lunacy. The plaintiff's counsel said, in court, "It was wonderful if plaintiff laboured under a dangerous lunacy, that no statute of lunacy had been taken out." The Attorney-General who appeared for the defendant asserted that, "If a verdict can pass against the defendant, madness must stalk at large through our streets, for no person will then keep a reputable private madhouse, and our public establishments will not be capable of holding the patients who shall apply." The judge, Lord Ellingborough, in his address to the jury, said that, "The defendant had justified, by saying, that he had taken the plaintiff on the representation of his friends, and that justification had been made out," but he did add, "There ought to have been another assignment on the record charging the defendant with confining the plaintiff longer than was necessary." Through an error in the indictment the plaintiff lost his case.

In another action the plaintiff alleged that his brother and another man had conspired to have him unlawfully, and maliciously, confined as a lunatic. A third man, the keeper of a madhouse at Hoxton, was indicted with having "unlawfully received the plaintiff into that house, the plaintiff being of sound mind at the time." The judge, Lord Kenyon, having heard in court that no medical certificate of insanity had authorized the plaintiff's detention said, that "he observed that part of the Act of Parliament respecting the reception of lunatics into places of this kind, namely that of having a certificate from a medical man, had not been complied with, and that of late years it was the custom not to be so strict in that particular." A verdict was given for the defendants, one of whom was Warburton.

These lawsuits will serve as a comparison to the case of Mrs. Mary Daintree, a former patient in Mrs. Burrows' madhouse at Hoxton, who alleged that she had been illegally detained in that institution. The certificate of insanity under which she had been confined was signed by James Parkinson, the visiting doctor. Three years and one month after she was released from her confinement, she prosecuted her nephew, his wife, and her own sister who, she alleged, had conspired together to have her put under restraint in a madhouse, though she was of sound mind at the time.

Unlike many of his contemporary medical practitioners, who had no conscience when it came to signing certificates of lunacy, Parkinson was an honest doctor, but was unfortunate enough to have been criticized adversely by the press and the public generally, for his testimony as a medical witness at the trial. He was accused of having certified Mary Daintree, not as a result of his own observations but in consequence of the opinions of others, the relatives in particular, who, it was alleged, had an ulterior motive in procuring Mrs. Daintree's confinement. He gave an account of the part he played in this deplorable episode in a tract published in 1811 (cf. infra).

In the extracts from newspaper reports of the trial given below, the account of Parkinson's evidence was noticeably brief and, since a thorough search of the newspapers of the time has not revealed additional details, it seems Parkinson's critics must have based their opinion solely upon this scanty information, unless they had attended the trial. This took place at the Sessions House, Clerkenwell on October 30, 1810, and lasted for seven hours. The most detailed report appeared in the *Statesman*, and was the subject of an editorial in the same issue. The headline: "Middlesex Sessions Conspiracy" would have attracted readers in search of sensational reading matter.

"Benjamin Elliot, Elizabeth Ann Elliot his wife, and Sarah Bodkin," The report began, "were charged on an indictment charging them with a conspiracy," the object of which was illegally to deprive Mary Daintree of her liberty (*The Statesman*, 1810).

Mr. Alley, with his usual eloquence entered into a detail of the case, which he observed charged the defendants, the first two only of whom appeared, with a crime of the most aggravated nature. It was no less than conspiring together for the purpose of accusing the Prosecutrix of being insane, and in consequence of such conspiracy actually confining her to a madhouse for fifteen weeks, from which she escaped by mere accident. The Prosecutrix, he stated, was left sole executrix of her husband's will, who died in 1806; and only guardian to her son, then a boy of about twelve or thirteen years of age. Her duty as executrix she duly performed of herself, until the period of her incarceration, regularly receiving her rents, and performing every other necessary function with the greatest regularity and minuteness. For the first twelve months after her husband's death she lived in her own house, but at the expiration of that time the defendant Elliot and his wife entered into the diabolical plot, which gave rise to the present prosecution. They commenced their atrocious machinations, by inviting the unfortunate Plaintiff to live in their house, under a pretence, that by residing by herself, her sorrows preyed upon her mind, and in fact, that for want of society and amusement, her days were passed in melancholy remembrance and useless regret; and, representing, if she came to live with them, they would comfort and solace her to the utmost of their power, and leave no means untried to render her situation desirable. In pursuance of their solicitations and promises, the unfortunate woman left her own house, and went to reside in the house of Elliot who was her nephew and who is a carver and gilder in St. John Street. But she had not long remained there before the wickedness of these persons was fully unfolded. For in the month of November, 1807, one night, while enjoying the pleasures of her own fireside, perfectly in her senses, and in sound health, the Prisoner Elliot, in pursuance of a previous concert with his wife, and Sarah Bodkin, the sister of the Prosecutrix, entered her room, accompanied by two keepers from a madhouse, threw her down with great violence, put her in a straight waistcoat, and conveyed her to Burrough's madhouse at Hoxton, where she remained for fifteen weeks. The atrocity of such conduct was too evident to require comment; and the object of the prisoners, that of possessing themselves of the property of the Complainant, was too clear to require further argument; he should therefore proceed to call the witnesses in support of the case.

A woman named Bailey, was the first person examined. She stated, that as she was passing the end of Burrough's madhouse, in the month of February, she was called by name from the garret window, by some person, who she recognized as

Mrs. Daintree. Several persons having been alarmed at seeing her converse with a supposed lunatic, assembled and asked what was the matter, but as she did not wish to excite any disturbance, she went away, and returned on the next day, when she was admitted to Mrs. Daintree, whom she had known for many years, and always conceived her to be perfectly regular in her senses and demeanour. Mrs. D. told her that she had been placed there by her cursed nephew and niece, who wished to enjoy her property. It was finally agreed between them that witness should exert herself to procure her discharge, which she was at length enabled to accomplish through the interference of friends.

Mrs. Wood, another friend of Mrs. Daintree, who was instrumental to the delivery, stated, that when she went to the house of Elliot she upbraided him with his inhumanity, and desired him to write to the brother of the Complainant, whose name is Surridge, and who lived at Epping; he said he had no objection to write to the brother, but that she should never leave the madhouse with his consent, on account of her former wickedness. She always thought Mrs. Daintree perfectly sensible and rational.

Mr. Wood, the husband of the last witness, deposed, that he wrote to Surridge, as the Prisoner Elliot seemed tardy in his motions, and described to him the situation of his sister. In consequence of which letter Mr. Surridge came to town, and after stating that he had been induced to give his consent to the confinement of his sister, from the representations of her nephew, Elliot and his wife, he obtained her liberty.

Mrs. Daintree, the Prosecutrix, was then called; she stated that she knew Elliot and his wife long before her husband's death, and that Sarah Bodkin was her own sister. About twelve months after her husband's death in compliance with the solicitation of the first two mentioned Defendants, she took a second floor in their house, but she had not long taken possession of her lodgings, when they evinced the greatest unkindness in their treatment of her, throwing every impediment in the way of her peace, and using every stratagem to render her life uncomfortable. She then corroborated the statement made by her Counsel, as to the manner of her delivery. During her examination she evinced the greatest propriety in her demeanour.

Mr. Alley here observed that he could bring a numerous host of witnesses to speak of her sanity, but he deemed it unnecessary as he was convinced the Court and Jury must see the justice of the prosecution.

Mr. Gleade, Counsel for the Defendants, next addressed the Court, with great feeling and ingenuity. He detailed various instances in which the insanity of the Prosecutrix was evident to all who had the opportunity of her acquaintance. He stated that she was frequently in the habit of fancying that persons were traducing her character, and even went so far as to accuse herself of being the author of her husband's death and to be in continual alarm lest the officers of justice should approach her for that offence. Sometimes she asserted that she heard whisperings in the air, that people spoke to her down the chimney, and that in fact, everyone was bent on her destruction. At other times; she was detected with cords in her pocket, and wrapped around her body or arms, which it was supposed she had concealed for the purpose of self destruction. On two or three occasions it was deemed necessary to watch her all night, and during that time she could not be induced to go to bed but walked about the room in the greatest disorder, until at length, from the increasing strength of her malady, it was thought advisable to remove her to her nephew's house, where he pursued her irregularities, and proceeded to such lengths at last, that the life of Elliot the Prisoner was endangered by a stroke which she gave him with a pitcher upon the head, during a dispute which originated upon a refusal to allow her to get water in the kitchen, after she had been abusing her niece.

It was not until this latter violence that she was removed to a madhouse, and on her arrival there, she had been declared by the visiting surgeon as a person whom it was highly proper to confine.

In support of these assertions the first person called was Mr. Parkinson, who stated himself to be the visiting surgeon of Burrough's madhouse at Hoxton: that he saw Mrs. Daintree in that house the morning after her conveyance thither; that from her appearance he thought she was not a fit subject for confinement, and therefore was resolved to make enquiries of the symptoms of her malady among her friends, and having seen her nephew and niece, and a lad introduced to him as her son; from their information he entertained no doubt of her insanity, and therefore declared her a fit subject for confinement.

Upon the son of the Prosecutrix being called, however, he said that he had always lived with his mother previous to her confinement, and that he never witnessed the slightest derangement in her manners.

Mr. Alley now addressed the Court, in reply, with considerable ability; and, by luminous argument, refuted the evidence adduced in favour of the prisoners.

The case being closed, the worthy Chairman observed, that from the nature of the testimony on both sides, it was difficult to decide which was correct. He did not conceive, however reprehensible the conduct of the female Prisoner might have been, that she ought to be included in the verdict of guilt against her husband, if such was the decision of the Jury; to them he would leave the facts, and he had no doubt but they would do justice to all parties.

The Jury, after a short consultation, returned a verdict of Guilty against Benjamin Elliot and acquitted his wife. (*The Statesman* 1810.]

Benjamin Elliot was sentenced to six months imprisonment in the House of Correction, Cold Bath Fields, Clerkenwell. *The Times* gave a very brief report of the case. The only witnesses mentioned were Parkinson and Mrs. Daintree's son, and their evidence was reported as follows,

The first of these witnesses was Mr. Parkinson, surgeon and apothecary at Hoxton, and medical attendant at the private madhouse where Mrs. Daintree was confined, and he proving his having certified for the necessity of her confinement, after a minute and circumspect examination. The son of Mrs. Daintree, who is in his clerkship with an attorney, was also called as an evidence for the defence; but swore that he had lived with his mother constantly since his father's death, and that he had seen her at the madhouse, but that he never observed any symptoms of derangement in her conduct (*The Times*, 1810).

The report of the trial in the *Morning Chronicle* consisted almost entirely of a dramatic description of the seizure of Mrs. Daintree and the mode of conveying her to the madhouse. Neither Parkinson's evidence nor that of any other witness was mentioned. The Prosecuting counsel's recital of Mrs. Daintree's ill-usage must have touched the hearts of the jurymen.

Under the heading, "Imprisonment in a Madhouse," the report proceeds to give an account of the manner of her treatment during the confinement.

Elliot held her hands, while his wife struck her several violent blows on the face. Not content with treating her in this manner, they encouraged her son to use her ill on every occasion. When she had been in their house near twelve months, she was sitting one evening in the month of November, at tea in her own apartment, when Elliott, the Defendant, came in, followed by two women, strangers to her, who she afterwards discovered were servants belonging to Mr. Burrough's private madhouse at Hoxton; one of the women produced a straight waistcoat, and was proceeding to put upon the prosecutrix, but she resisted, upon which the other woman named Luxton, seized her by the hair and pulled her down on the floor, when with the assistance of Elliott and the other woman, they succeeded in putting the waistcoat upon her; then they took her up, the women by the arms and Elliott by her feet, and carrying her downstairs, put her into a coach, when Elliott and the women got in with her, and Elliott told her they were going to take her to her brother's. On the coach stopping however, she discovered that she was at the gate of Burrough's madhouse at Hoxton and she told them so, on which they confessed she was right.

On her way from her own house to the madhouse she had frequently attempted to speak and tell Elliott she knew the property she had left behind, but was continually prevented by his putting a handkerchief before her mouth. On their arrival at the madhouse, she was taken to a parlour where Mrs. Burroughs the mistress of the house came soon afterwards. Luxton came and took from her keys, money, and whatever else she had in her pocket. She was then taken upstairs to her bed, which was extremely narrow and uncomfortable. She begged to have the waistcoat taken off, but she was told she must wear it during the night. It was, however, taken off next morning, and not put on any more during the time she was in the house. She told Mrs. Burroughs frequently while she was in the house, and represented to her how ill she had been treated and begged to be liberated but was constantly refused. When she was taken to the madhouse by Elliott she left upwards of twenty pounds in cash in her apartment, and several other things; none of which were to be found on her return there.

The defence attempted to be set up was, that the Prosecutrix had been deranged, and that it was necessary to confine her to prevent her doing herself an injury. Several witnesses were called to prove this fact, but their testimony was extremely vague and contradictory.

It appeared, however, that although Mrs. Elliott had been guilty of cruelty and ingratitude towards the Prosecutrix, yet that she had not taken an active part in conveying her to the madhouse, or accompanied her in the coach when she went there. The Jury therefore acquitted her but found her husband Guilty (*Morning Chronicle*, 1810).

A few weeks previous to the trial, the *Statesman* had published a letter describing a gentleman who, while confined in a private madhouse, was "chained round his neck, waist, and legs, and beat unmercifully with a large stick by a ruffian who was hired for that purpose; and that he was half starved." The writer suggested that Parliament should immediately pass an act incorporating new regulations in that no person should be confined before being examined by three practising physicians; that the madhouses should be visited every day by a doctor; that a Board of Physicians should examine and report on each inmate every three months; that the worse cases should not be treated in the same house as the milder ones; and that all private madhouses should be converted into hospitals, with gardens and grounds attached, for exercise and recreation (*The Statesman*, 1810a).

The Statesman attacked Parkinson's conduct in certifying Mrs. Daintree. Its editorial stated:

In a subsequent column, we have given an interesting trial of yesterday, at the Middlesex Sessions. Benjamin Elliott, and his wife, and Sarah Bodkin, had obtained the confinement, in a madhouse at Hoxton, of Mrs. Mary Daintree, on a plea of that Lady's insanity. They were relations of Mrs. Daintree; and it was alleged that they had sent her to the madhouse, in order to possess themselves of her property. There is much contradiction in the respective evidence for and against the prosecution. But it is to be observed, that where people are really wicked enough to commit the enormous crime with which the Prisoners were charged, it is easy for them to procure witnesses, in this motley metropolis, to make out their case, according to the instructions of unprincipled advisers. We cannot say this was the fact here; but it is plain that the evidence for the prisoners was not believed by the Jury, who found the man Elliott, *guilty*, although they acquitted his wife. *Sarah Bodkin* did not appear. The conduct of Mr. Parkinson who stated himself to be visiting surgeon of the Mad-house, however, deserves public attention, both as it stands in itself and respects himself, and it loudly calls for *public interference* in the private mad-houses. He was that he saw Mrs. Daintree in that house the morning after her conveyance thither; that from her appearance he thought she was not a fit subject for confinement, and therefore was resolved to make enquiries of the symptoms of her malady among her friends, and having seen her nephew and niece, and a lad introduced to him as her son; from their information he entertained no doubt of her insanity, and therefore declared her a fit subject for confinement.

How did Mr. Parkinson come to this conclusion? He did not *himself* think the lady insane, but he believes her to be so, on the representation of relations, whom he might have suspected to be interested in her confinement! And is the deprivation of people's liberty and peace of mind, is the infliction of torture upon them, to be a matter put at the discretion of a gentleman who could argue as Mr. Parkinson did in this case? It is for the purpose of introducing this *most* important question to the public, that we have made these brief remarks on the trial; and we shall take an *early* opportunity to resume the subject. (*The Statesman*, 1810b).

The editor, however, did not return to the subject of madhouses and the certification of lunatics, but two weeks later he published a letter from a correspondent making a further but more ferocious attack on Parkinson. The letter, dated October 31, 1810, appeared in the issue of November 14, 1810.

Sir,

I have read last night your report of the trial, Daintree v. Elliot and Others. That cause is a very interesting one; and of such a sort *of interest* that every individual feels more or less, a party in the discussion.

The evidence of Mr. Parkinson is *very extraordinary*, and places him not in the most amiable point of view, there is, however, at all times, an odium attached to every medical man in the *regular employment of these Houses*. I would like to ask Mr. Parkinson who made him *Visiting Surgeon* of that Madhouse; by whom was he appointed; by whom can he be removed; and whether he has not a share, or some other *interest* in the profits of that house; or of some other of that description in his neighbourhood? It is the duty of every medical man, when he is called to visit a person within those walls, to ascertain the *legality*, and the necessity of the certificate, signed by some persons, one of whom must be either a physician, surgeon, or medical practitioners of reputation. With this detail Mr. Parkinson *must* be familiar. Let me ask him, then, whether he ever saw *this* certificate, and let him produce a copy of the signature to it? If he is the '*regular visiting surgeon*' he can be at no loss for this. If there is no such certificate the keeper of such house is punishable.

I highly commend you for the observation with which your paper noticed this infamous transaction, and I hope you will still afford your assistance to a liberal discussion of this lamentably neglected subject.

To the solicitor of Mrs. Daintree I would advise ascertaining whether the Certificate of Admission was *legally* signed, and by whom—if *real* or forged signatures—whether Mr. Parkinson did not himself sign the Certificate—who are the parties in that house—(whether they are enabled to play into each other's hands and elude the inquirer after *lost persons*, decoyed from the basest motives.

Who are the parties in that house—whether the keeper there, and those at Bethnal Green are not the same, so that they are enabled to play into each others hands and elude the inquirer after *lost persons*, decoyed from the basest motives. After he has sifted this affair minutely, let him face the public, with the details of the trial, punishment, and all the facts illustrative of this transaction; in the doing of which, he will be rendering the public an essential service.

There is a Society for the Suppression of Vice, & : how much would it add to their usefulness if they would take under their notice these houses, and the conduct of those who are interested in them: if *they*, or a distinct set of friends in the cause of humanity, a cause more talked about than acted upon, I will add my mite and my assistance as far as my abilities are capable.

It has fallen to my lot to become acquainted with some facts in respect of Private Mad-Houses, which I shall watch for some favourable opportunity of using for the betterment of these establishments, or for their total change of system, if not their annihilation altogether, as soon as the system for re-building Bethlem is in earnest commenced—I mean within the Bills of Mortality. If this idea is not practicable, that there should be *no private house allowed, but under the eye of a deputation from the Governors of St. Luke's and Behtlem Hospitals*; for I believe the visitors from the College of Physicians have hitherto done no *services*. The intention is well—but it is completely defeated. Is it reasonable that physicians in any degree of practice can be expected to act gratuitously upon such a painful duty—that is expecting too much from men *so situated*? The real services that are expected are defeated; but on this *particular* point, I may address you more in detail at some future period.

I am, Sir,

Your humble servant,

M.E.N.S.

Oct. 31, 1810 (*The Statesman*, 1810c).

"In one or two of the newspapers of the next morning," wrote Parkinson in the prefatory address in his own tract, "a very incorrect account appeared of the evidence delivered on the trial, and particularly of that which I had been called upon to give."

On these erroneous statements, reprehensive observations were founded, and were published in one of the newspapers of the following evening. The censures which these conveyed, were echoed and reechoed until my friends requested that they should be seriously noticed. This I conceived by no means to be necessary, trusting that no one who knew me would give credit to my having committed myself, by conduct so reprehensible as that which had been imputed to me. At the end of a fortnight, a letter on the same subject, too contemptible and illiberal to be answered, appeared in the same paper . . .

The next attack was from a short lived medical newspaper, The Appollo, and as I am informed, with a considerable degree of contemptible malignity. Of this, however, I do not speak on my own knowledge, for, on endeavouring to obtain a copy of the paper, the publication was found to be stopped. Still it did not appear to be necessary to obtrude myself on the public on account of these attacks.

At length I obtained information that the visitors of the college had expressed their surprise that I had not contradicted the calumnies which had been raised against me. On hearing this, I became somewhat disposed to correct the misstatement of my evidence, which had appeared in the public papers, but I did not resolve on undertaking the easy task, until I had led myself to hope, that, at the same time I might offer some observations which might show the defects of the act for regulating mad-houses, with some incidental remarks, which might be useful to those who had relatives suffering under the dreadful loss of reason. With these hopes, the following pages are humbly offered to the public.

The footnote states,

As a charge of illiberality, although against an anonymous writer, should be supported by evidence, the following passage is offered for this purpose, and, it is supposed, must prove satisfactory. Who are the parties in that house? Whether the keeper there and those at Bethnal-green, are not the same, so that they are enabled to play into each other's hands, and elude the inquirer after *lost persons*, or persons decoyed from the *basest motives*?

I cannot help here observing, in justice to Mrs. Burrowes, that I believe no other party has any concern in the house which she keeps; that she has no connection with any other house of the same description, that her character is unimpeachable.

With equally good information, this writer observes, that 'he believes the visitors from the college of physicians have hitherto done *no services*.' There is one other person who entertains, or did affect to entertain, similar notions with those of this writer. Mr. Alley promised the jury on this trial, that the business should not stop here, that however high their situation, these gentlemen should be taught that they were not to wear their honours like useless nodding plumes (Parkinson, 1811a; V–VII).

The tract was written and completed by January 1, 1811, so only two months elapsed between the trial and Parkinson's reply. In it, Parkinson gives his own version of the events which led to the trial.

Nearly four years ago, Mrs. Mary Daintree, who had lost her husband some months before, became so much deranged in her mind as to be troublesome, by the strangeness of her conduct, to the neighbourhood of Rosoman Street, Clerkenwell, in which she resided. She was sometimes seen walking up and down the street where she lived, at four or five o'clock in the morning, in the most agitated manner. Sometimes she fancied that voices came down the chimney, or through the wainscot, or were to be heard in the air, calling her reproachful names, and accusing her of killing her husband. At other times, she would be seen walking about with a long cord wrapped many times round her body, or her arm, saying that she had been ordered to wear it, and would not admit of its removal. One evening, in particular, her distress seemed to be so great, and she having tried to conceal a cord and then a knife, and it was supposed for the purposes of self destruction, that it was found necessary to sit up with her all the night. Seeing her in this state, an intimate friend made application to her brother, and it was concluded that she should be gotten into St. Luke's Hospital; but on getting the blank tickets, and finding that, as the relations thought, her having an income of eighty pounds per annum would render her unfit to be offered as an object of charity, the tickets were not employed.

At this period, she repeatedly applied to Mr. and Mrs. Elliot, her nephew and niece, and requested that they would let her have an apartment in their house for herself and son; complaining that the wretches would never let her alone in Rosoman Street; that they were always calling her names, and charging her with being the death of her husband; that the hawkers of coals went about there with a bell, giving out that she was a bad woman; and that the children were set on by the wretches to abuse her.

Mr. and Mrs. Elliot unfortunately complied with her request; but she soon became so exceedingly noisy and mischievous, still fancying that her tormentors followed her with their voices, and abused her, that application was again made to her brother, who then desired they would get her into a private mad-house, which, as she became exceedingly violent and dangerous, was done on the 28 of November 1807.

On the following morning I first saw her, at the desire of Mrs. Burrowes, but finding her answers to be rational, although her manner was strange and eccentric, I declined signing the certificate until I had gained more information respecting the nature of her insanity. Having inquired therefore from where she was brought, I went to St. John's Street, and wishing to obtain my information from those who were least likely to be interested, I first of all made inquiries at the house on each side. At one of the houses I was informed that they knew little of her, but that their lodgers were often disturbed by the noises which she made. At the other house, I believe neither the master nor mistress was at home; but as to this I am not sure, my memory not serving me correctly, it being full three years since.

I then went to the house, and saw Mr. and Mrs. Elliot, but told them on their proceeding to give me their reasons for supposing her insane, that I must have other evidence, if possible, than theirs. I therefore begged to know who else lived in the house, and was informed, a lodger, and Mrs. Daintree's son, both of whom I had called to me.

Upon inquiring of the lodger, she declared that she was convinced of her insanity, that she had heard her complain of hearing voices; that as she lived in the room over her, she could frequently hear her, when she knew her to be alone, talking loudly, as if in conversation with persons in her apartment; and that she had seen many instances of her violent and dangerous conduct. I then saw her son, a youth of about 16 years of age, and inquired of him what he thought of his mother's situation, to which he answered, that she was certainly mad and upon someone observing that before this she had been rather in a low desponding way, I asked the son if he could form an opinion as to the cause; but the answer which I received, I refrain from giving here, from motives of delicacy.

Thus furnished with the necessary points on which to examine my patient, I returned to her, and after some further conversation, was fully convinced of her being a lunatic, and signed the certificate.

Her confinement lasted about three months, during which she was visited by her son every week or fortnight, at the importunity of Mr. and Mrs. Elliot, with whom he continued to live. At this period she saw an acquaintance from one of the windows, whom she called to, and informed that she was confined there though perfectly in her senses, and desired she would come and see her. Her acquaintance, therefore, applied to Mrs. Burrowes, who immediately introduced her to Mrs. Daintree, and upon her brother's coming to town she was released, in consequence of the repeated applications of these acquaintances.

On October the 29th 1810, three years within a month after the time of signing the certificate, I received a subpoena, and attended as a witness at the trial. . . . On being called up, the following examination took place.

Q. When did you see Mrs. Daintree first at Mrs. Burrowes?
A. The morning after she had been brought there; when I was surprised at finding her the widow of a man whom I had known and respected.
Q. In what state did you find her?
A. She seemed well in health, and her conversation appeared to be rational, but her manner was strange and peculiar. I would wish to observe that I was then unable to determine, whether that was the consequence of irritation or not.
Q. Well, sir, what then?
A. I acquainted Mrs. B. that I did not see sufficient reason to convince me of the propriety of signing the certificate; but that I would call in St. John's Street, and endeavour to inform myself as to the nature of her madness.
Q. Well, sir, you then went to St. John's Street?
A. Yes, sir.
Q. Well, Sir?

Mr. Alley, addressing Mr. Gleed in an under voice, said We'll have none of what passed in St. John's Street. Mr. G. Why not? Mr. A. Because it is conversation.

Mr. G. Well then Mr. P. you must not give us conversation—you must not tell us what passed in St. John's Street—only tell the result—did the inquiries you made induce you to sign the certificate?

A. The information I obtained, induced me to return to the house, and *again examine the patient*. Having done this, I was satisfied of her being a lunatic, and of course signed her certificate. Mrs. Daintree's son being examined, he swore 'he had never seen her any more mad that she was then.' On this, I stated in a note to Mr. Gleed, the counsel for the defendant, that I had examined the son of Mrs. Daintree before I signed the certificate, and that he had even given a cause for it. Mr. Gleed then asked him if he remembered Mr. Parkinson examining him respecting the state of his mothers mind? He answered, No. Then, said Mr. Gleed, I need hardly

ask you if you acknowledged your mother to be out of her mind. To this he replied, Never. I was then again examined by Mr. Gleed.
Q. Who did you see at St. John's Street?
A. A youth, who I understood to be the son of Mrs. Daintree; but the court will be pleased to observe, that at three years distance, I do not swear that the youth I then saw is the same who has just given his evidence.* Mr. Gleed was proceeding with his examination, but was opposed by Mr. Alley, on the ground, that he was endeavouring to impeach the veracity of his own witness (Mrs. Daintree's son).

This unexpected deposition, from one of the witnesses which had been called by the defendant, appeared to be the circumstance which chiefly influenced the jury in their verdict which they gave against him, and which was followed by a sentence of six months imprisonment in the house of Correction in Coldbath-fields.

A reference to the report of the evidence as it was actually given (P. 43, &.) will show that the report commented on by the editor of the Statesman, is exceedingly erroneous in its most material points, those which are distinguished by being printed in Italics.

Instead of deposing that 'I thought she was not a fit subject for confinement,' I swore that her manner was strange and peculiar—that I did not see sufficient reason to convince me of the propriety of signing the certificate; but that I said, that I would call in St. John's Street, and endeavour to inform myself of the nature of her madness. And instead of stating that having seen her nephew and niece, and a lad who was introduced to me as her son, that from their information I entertained no doubt of her insanity; and that I *therefore* declared her to be a fit object of confinement, it will be seen that the information I obtained, induced me to return to the house and *again examine the patient, and that having done this, I was satisfied of her being a lunatic, and consequently signed the certificate.*

The statement of the evidence being thus shown to be erroneous; no necessity exists for refuting the charges, which are implied in the commentary of the Statesman, on the evidence thus reported. Indeed my constant answer has been to those who wished me to notice the remarks of the Statesman, that I had no fault to find with the editor, but his too easy confidence in the report of the evidence; and that the remarks which he had made were such as he would have been fully warranted in making, had the report of the trial been correct.

Convinced as I am, that Mrs. Daintree was actually a dangerous lunatic, at the time of her admission into Mrs. Burrowes'—doubting even her cure being complete when she was liberated—not being able to ascertain that the nephew Elliot could derive any possible advantage from her confinement—and knowing that it was with the wish of all her near relations, and under my certificate, granted after a careful scrutiny, that her confinement took place, I cannot but lament the sentence which was passed. I must indeed go further, and encourage a hope that the unfortunate prisoner will, on his return to society, find that his former connections will cheeringly continue to him that support which has been hitherto yielded to him, as an unexceptionable, and as I am informed, a most meritorious character.

Parkinson made no comment on the insinuation made by the anonymous correspondent that a medical certificate was necessary before the

---

*Recollection however, soon convinced me that it was the same young man (Parkinson, 1811a; 39–48).

admission of Mrs. Daintree to the madhouse, and that it had either been deliberately withheld, or lost. Had it been received it would not have been necessary for Parkinson to issue another certificate. He was within the law in signing such a certificate on the day following admission, since it could be deposited with the beadle of the Royal College of Physicians, for the attention of the Commissioners in Lunacy, within three days of the date of admission. Certification generally was perfunctory. In the lawsuits quoted above, Chawner v. Warburton, and the case concerning an unnamed plaintiff who was an inmate of a Hoxton madhouse, the judges, as will have been seen, either made no comment on the absence of a certificate or merely observed that they were sorry to hear that the law had not been observed in that respect. Dr. Richard Powell (1767–1834), physician to St. Bartholomew's Hospital and secretary to the Commissioners in Lunacy, said some four years later that he could only recall one case in which a prosecution was instituted by the Commissioners against the keeper of a private madhouse for admitting a patient without a certificate. Mrs. Daintree's certificate was signed, not by an ordinary general practitioner, but by one who from his vast experience was an expert in insanity.

The result of the trial of Mary Daintree, and the attacks on Parkinson which appeared in the daily newspapers, must have harmed his professional reputation as well as his personal esteem. Nevertheless, he used the tract not simply to vindicate his own actions, but to reflect on the law regulating private madhouses.

Parkinson applauded the passage of the 1774 Act, "the objects of which are, the regulation of mad-houses, the admission of patients, and the visitation of these houses and patients by proper persons." Evils, however, still remained, some connected with the state of medical education. Physicians and surgeons, he insisted, should be able to diagnose madness.

CHAPTER 11

Mr. Parkinson, the Palaeontologist

To James Parkinson, fame came in the medical world posthumously by his identification of the disease which he termed "The Shaking Palsy," paralysis agitans. Fame came to him in his lifetime, however, in that branch of geology called Palaeontology, but known in his day as Oryctology. In Chapter 2 we mentioned his celebrated museum at No. 1 Hoxton Square, with its valuable collection of fossils. In those days, there was keen competition for fine fossils, and Parkinson is said to have paid as much as 20 guineas for a specimen. An insatiable collector, he included not only fossils, but minerals and shells, delighting in the beauty of their form and colour. His "cabinet of coins and medals" was bequeathed after his death to his brother-in-law, John Keys.

Medical men in the 17th and 18th centuries, by virtue of their scientific scholarship, were particularly attracted to a study of the Earth and its fossils. Parkinson was a worthy descendant of a long line of British medical men whose researches and writings contributed so much to the creation of the science of geology, namely Martin Lister (1639–1712), John Woodward (1665–1728), Robert Plot (1640–1969), Charles Leigh (1662–1701?), John Beaumont (?–1731), Samuel Dale (1659–1739), James Hutton (1726–1797), and John Hunter (1728–1793).

"Mr. Parkinson, the Palaeontologist," as he was referred to in the middle of the 19th century and afterwards, would have described himself as an oryctologist or fossilist, since the term palaeontology was not heard of until 10 years after his death, when Baron Fischer de Waldheim (1771–1853) first suggested that it was more accurate and more descriptive than the older term, oryctology.

Parkinson began to study fossils at a time when "geology" as a distinct science was just beginning to appear. English naturalists straddling the turn of the 17th–18th centuries had produced a bewildering number of theories of the Earth, many of which were based on contemporary physical science and Bible study, and most provoking refutations and counterarguments. After a quieter period of consolidation in the middle of the 18th century, another wave of theories appeared in the 1780s and 1790s.

These were based on the more sophisticated physical science of the day, but were still largely theorizing exercises.

The French Revolution made such works seem dangerous, and practitioners of geology, as the study of the Earth was now called, were forced to take up a more empirical position. The period 1790 to 1820, once known as "The Heroic Age of Geology", marks the period of detailed studies of rocks, fossils, and minerals, at first in their own right, but later in relationship to the strata in which they are found. It was during this period that a consensus of opinion emerged that the Earth had a long and complex history, had undergone certain progressive changes during its history, and that the distribution of strata and their fossils needed further study.

It was a copy of *Chemical Essays* (1781–87) by Richard Watson (1737–1816) that first turned Parkinson's attention to chemistry and introduced him to geological subjects. He acquired his knowledge of fossils by collecting specimens in the field, dug out with the traditional geological hammer. Dealers in fossils advertised that they gave instruction in "mineralogy and oryctology," thus encouraging the study of those branches of geology at the same time enlarging their sales of specimens. Parkinson's earliest fossils (sometimes termed "Medals of Creation") were dug out, either by himself or by the workmen, in the gravel and clay pits of Shoreditch or West Hackney, and in particular from "a tile kiln and brickworks" owned by William Rhodes (an ancestor of Cecil Rhodes) at Haggerston, in the parish of St. Leonard's, Shoreditch. He later enlarged his collection by exchanging duplicates, purchasing from dealers or at auction sales. Parkinson's copies of 18 natural history sale catalogues are preserved in the Zoology Library of the British Museum (Natural History), dating from June 1798 to May 1808, and give a good idea of his tastes and interests.

In the preface to the first volume of his book, *Organic Remains of a Former World* (1804–11), Parkinson says that "Impelled by that eager curiosity which a view of the remains of a former world must excite in every inquisitive mind, the writer . . . long and earnestly sought for information respecting the wonderful substances, from every source to which he could obtain access."

By temperament modest, he was hesitant about publishing a book on fossils though he had long contemplated doing so. He wrote: "The kind encouragement of his scientific friends prompted him to apply himself to the task." But there were others with opposite views, which fact "almost induced him to relinquish his proposed intentions" by those with wide experience of publishing, who said frankly that there was but "a small demand" for a technical work which could only be of interest to a small number of readers. Despite this discouragement, "a conviction that ample support might be obtained, were the attention of the public fairly called to the wonders which the study of oryctology discovers, made him resolve the attempt." He was well aware that the number of collectors of the

"Medals of Creation" was small and he feared that "a dry, strictly scientific work" would not appeal to "readers in general". With a desire that the work might find favour with the general public, he resolved to adapt it to the intelligence of the ordinary readers whom he hoped to induce "to the study of this part of Nature's works, to whom, as yet, they might be unknown." With this in mind, he decided to write in "the epistolary form, from the familiarity of style which it permits." Letters were to replace the usual enumerated chapters, letters addressed to an imaginary friend, a learner with an inquiring mind. By adopting this method, he hoped "the matter . . . will be found highly interesting" and "sufficient to awaken attention to a most pleasing but much neglected science" (Parkinson, 1804–11, Vol. I, p. VI–VII).

To anyone familiar with the occupation of a general medical practitioner, working single-handed in a long-established medical practice in a densely populated district, it is difficult to conceive how Parkinson could find leisure for such a time-consuming task. "These observations," he wrote, "have been put to paper, at short and uncertain periods" and when his mind was often fatigued from professional labours.

His eldest son John, who wished to become a surgeon like his father and grandfather, was apprenticed to his father in the year 1802, and it can be assumed that he was able to relieve his father of duties in the apothecary shop, compounding medicines and dressing wounds just at the time when James Parkinson settled down to the preparation of the first volume of his work. In July 1804 the first of three quarto volumes, which were entitled *Organic Remains of a Former World. An Examination of the mineralized remains of the vegetables and animals of the antediluvian world generally termed extraneous fossils* was published. It was printed at the author's expense and the complete text, printed on quarto sized paper, covered 1146 pages and was accompanied by some 42 plates containing 700 figures, nearly all of which were drawn from specimens in Parkinson's own collection and were remarkable for accuracy and realism. *Organic Remains* continued unchallenged as the standard textbook on Palaentology for half a century, and its illustrations were reproduced as late as the year 1850.

Organic Remains was intended to be essentially a practical treatise on fossils which would assist the student of oryctology, digging for himself, to identify his discoveries.

A most interesting and informative history of the science of oryctology, extending back to the Greek and Roman philosophers who wrote of the discovery of petrified shells in the bowels of the earth and on the tops of mountains, is given in Letter III of the book. They expressed astonishment and bewilderment as to how the shells came to rest in these situations so remote from the sea. Pliny described certain petrified substances which he named according to their resemblance to familiar objects; e.g., Bucardia, like an oxen's heart; Hammites, like the spawn of fishes; Ostracites, like oysters; Spongites, like sponges, etc. However, these names did not mean that Pliny gave any opinion as to their origin.

Parkinson, in asserting that fossils are the remains of animals and plants destroyed by Noah's Flood, is almost the last in a line stretching back to the 16th century. Initially, the flood was used as a simple explanation of the distribution of fossils over hills and mountains far from the sea. In the 17th century naturalists had difficulty reconciling the details of the Bible story with the sort of flood required to explain the fossils. In the 18th century, knowledge began to accumulate on the occurrence of different fossils in different rock layers and on the regular arrangement of rock strata. In 1804, Parkinson appeared to have been blissfully unaware of these difficulties, and was content that Noah's Flood would explain everything.

The author says (Letter XXIX) that his theory of the petrification of wood differs from that of the "first authorities." He rejects the theory of the absorption of the organic material and its substitution particle by particle by siliceous or calcareous substances deposited from the surrounding water which had held them in solution, and he also rejects the theory of James Hutton that siliceous wood is the result of "injection with flint, in a state of simple fusion by fire." He believes that the wood is thoroughly soaked in water which is "saturated with earthy particles, which are finally consolidated by crystallisation at once infixing in the stony mass thus formed, the undisturbed arrangement of the enclosed vegetable fibres." William Martin disagreed with this hypothesis and observed: "On the whole we see no reason to reject a theory so generally received for that which Mr. Parkinson has endeavoured to introduce. At the same time we beg to state the greatest respect we entertain for that gentleman's chemical knowledge, as well as for the ingenuity displayed in the investigations of the present subject" (Parkinson, 1804–11, Vol. 1 p. 309).

Palaeobotany, the science of fossil botany, had hardly come into existence when Parkinson first began his study. Coal mining had first drawn the attention of the scientific world to fossil plants during the beginning of the 18th century. Plants generally were too delicate for fossilisation; trees, ferns, and fruits, however, were plentiful. Parkinson found "great difficulty in ascertaining even the genera of plants," since the majority found fossilised "have been acknowledged to belong to plants not known to botanists; none resembled any existing species." The extreme frailty of flowers, Parkinson maintained, accounted for their absence in the fossilised state. As earlier observers had been skeptical as to the vegetable origin of fossil plants, so later enquirers went to the opposite extreme. They used their imagination freely and saw in imperfect impressions "ears of corn and maize, flowers of the tulip and the aster, and the fruit of the pineapple!" He attributed the discovery of petrified fruits to "freaks to nature." "I can assure you," he says, "I could have laid before you the figures of apples, pears, plums, &c., which would pass with many unquestioned; but which undoubtedly owe their form to mere accident." "Of all the sports of nature," he remarks, "in the formation of stones, these are among the most delusive and unaccountable: judging from a specimen in my possession, which I understand was purchased by the late Mr. Strange, from Germany at considerable trouble and expense. Its form is that of a long square, being similar to what are called Dutch loaves" (Parkinson, 1804–11, Vol. I, pp. 449, 463).

He concludes his final letter with some reflections on the subject of the Deluge. "Why the earth," he reflects, "was at first so constituted that the deluge should be rendered necessary—why on earth could not have been at first stored, with all these substances, and endued with all those properties, which seem to have proceeded from the deluge—why so many beings were created, as it appears, for the purpose of being destroyed—are questions which I presume not to answer. Trusting, however, that what has already been said must render their solution less difficult, and lead to conclusions less repugnant to reason, than some which have been advanced, by men of considerable abilities, and learning, I shall here close our correspondence, for the present, experiencing no small degree of satisfaction at finding every fact I have noted, and every conclusion they have led me to make, to be in perfect accordance with the highest sentiments which can be formed respecting those laws, by which the regulation of the economy of creation was decreed" (Parkinson, 1804–11, Vol. I, p. 471).

The second volume of *Organic Remains* is entitled "The Fossil Zoophytes." In the preface, Parkinson says that he regretted that his treatise could not be completed in the present volume, nor could he say how many volumes would be needed for its completion. He preferred to provide figures of original drawings made from specially chosen specimens of his own collection, rather than copy them from the compilation of others, "so much of the labours of preceding writers," he considers, "as should serve to furnish a correct history of the progress of knowledge, respecting each subject of examination" would be given. Numerous difficulties had confronted him, and the greatest difficulty appeared to be "the acquisition of a sufficient variety of specimens, to admit of the necessary comparison, and to allow for the deficiencies of some, so to be made up by others, as to supply a tolerably correct notion of the figure, and even of the nature of the body, which perhaps no single specimen could furnish." Broken, incomplete, or partial specimens must have tried his patience and ingenuity (Parkinson, 1804–11, Vol. II, pp. XI,XII).

In defending his preference for a continuation of his volume in the form of letters, he maintains that "The epistolary style too appears to be well suited to the discussion of subjects of natural history. It allows that familiar form of expression, which seems to be best calculated for those illustrations which are required by the various topics which the study embraces. It also admits of the introduction of those reflections which, although perhaps not unacceptable, and which naturally arise from the immediate subject of inquiry, might appear too digressive, in a work written in a truly systematic form. But another circumstance, it must be admitted, has had some considerable weight in determining this choice. These observations have been put to paper, at short and uncertain periods, when the fatigue arising from professional labours have left the mind in such a state, as to render it necessary to record the results of investigations, in that familiar kind of style as might best suit the writer in such moments; and might not demand the extreme severity of criticism" (Parkinson, 1804–11, Vol. II, p. 2).

Parkinson's muriatic acid test was a discovery which attracted considerable attention in scientific journals of the period. He was able to show the presence of organic matter in marble which consisted of animal fossils embedded in limestone. A piece of marble was suspended in a beaker containing a dilute solution of muriatic acid, and what occurred subsequently he describes in these words: "as the calcareous earth dissolved, and the carbonic acid escaped, I was much pleased, to observe the membranaceous substance appear, depending from the marble, in light, flocculent, elastic membranes". "Thus it was proved," he said, that the fossilising matter "was deposited in and around the animal structure." Plate 1, Fig. 1, in Volume II illustrates the result of this experiment (Parkinson, 1804–11, Vol. II, p. 16).

The clay and gravel pits of Shoreditch and Hackney were a happy hunting ground for Parkinson in his early days and indeed they seemed to provide an inexhaustible source of specimens. "In some of my earliest researches", he said, "in the gravel pits in the neighbourhood of the metropolis, my attention had been repeatedly attracted by a certain species of pebbles, which, from their general uniformity of character, I was convinced, must have owed their form to either vegetable or animal organisation. I therefore collected a considerable number of them, hoping that, by frequently repeated examination and comparison, I might be enabled to form some probable opinion respecting their origin: my endeavours, however, were by no means productive of that satisfaction which I wished.

Anxious to obtain what knowledge I was able respecting the formation of these pebbles, I broke many of them in different directions, and was thereby enabled to ascertain, that their shape and their openings depended on the figure of a body, which they contained in a cavity from some lost species of Alcyonium or sponge, I am fairly of opinion; and trust that in this opinion I shall be joined with those who are acquainted with the characters of these two genera." Time and his successors have proved the truth of his observations on the structure of a pebble (Parkinson, 1804–11, Vol. II, pp. 110–11).

The third and last volume of *Organic Remains*, published in 1811, was subtitled "Fossil starfish, Echini, Shells, Insects, Amphibia, Mammalia, &c." and contains 455 pages divided into 32 letters, together with 22 plates and 359 figures. The frontispiece illustrated "The back grinding teeth of the Mammoth or Mastodon of Ohio, Weight 4 lb. 7 oz., drawn by Springsguth from the original specimen" (Parkinson, 1804–11).

In the preface, there is a reference to the recent "important observations of M. Cuvier and M. Lamarck," which are incorporated into the present volume. He now had the advantage of Lamarck's classification of fossil shells and Cuvier's work on the fossils of the higher animals; England and France were at war, but the books that supplied this information were nevertheless freely available in Britain. Parkinson acknowledged them both:

Excepting the *Fossil Hantoniensia* of Solander and Brander (states the preface), no really systematic arrangement of fossil shells had appeared; the classification of shells, therefore, by Lamarck, in which particular attention is paid to those in a fossil state, became highly estimable. So clear and comprehensive is the arrangement of this naturalist, that of the numerous fossil shells which were unclassed, there were hardly any which may not now be placed under an appropriate genus . . .

Very few successful anatomical examinations of the fossil remains of *amphibia*, and of land animals, had been attempted before the justly celebrated Cuvier made them the subject of his investigation; but in consequence of the ardour with which he has availed himself of the extraordinary opportunities which he possessed, the history of these fossils must now be chiefly formed with the materials which he has furnished

(Parkinson, 1804–11, Vol. III, pp. IX–X).

Parkinson had avidly read Cuvier's "Leçons d'Anatomie Comparée," published in 1805. It must also have encouraged him to attend the lectures in Comparative Anatomy, given at St. Bartholomew's Hospital by Dr. James Macartney (d. 1843), at a period just previous to the publication of the third volume of *Organic Remains*. Fully a third of this volume covers the field of vertebrate palaeontology, the fossil remains of fish, birds, amphibia, reptiles, and mammals.

Cuvier (1769–1832) and Lamarck (1744–1829) are acknowledged to be the founders of vertebrate and invertebrate palaentology, respectively. Parkinson dutifully follows Cuvier's classification of the vertebrates which had been adapted from that of A.H.A. Dumeril (1812–1870). He regretfully departs from the Linnean tradition when he says: "I must here also crave your attention, which I excuse myself for departing from that classification which has been so long established by the truly great Linnaeus" (Parkinson, 1804–11). He explains his seemingly unorthodox act of renunciating such a firmly established system:

It must, I think (he contends), be admitted, that many of his genera are too comprehensive; added to which, since the forming of his classification, several shells have been discovered, possessing characters which will not allow them to be placed under any of his genera. The latter circumstance had indeed taken place to such an extent with respect to fossil shells, as to have rendered the formation of a new classification absolutely necessary.

This task the celebrated Lamarck has accomplished, with such ingenuity and care, as to give him a strong claim on the gratitude of every lover of science: and, as was exceedingly desirable, he has, by his classification, secured the admission of all these shells, which are found in a mineralized state

(Parkinson, 1804–11, Vol. III, pp. 47–8).

Parkinson was interested not only in fossil shells but also in recent shells, of which he had a large collection. He must have been considered an expert conchologist, since he wrote the article on shells in Nicholson's *The British Encyclopaedia* (1809).

The fossilized remains of "sea-urchins, sea-hedgehogs, sea-thistles," are dealt with in Letters II–IV, under the heading "Echinites," the mineralized remains of the echinus or sea-urchin. Here again he departs from the Linnean classification.

In his last years, Parkinson continued his researches into the mechanism of the siphuncle. His theory, that in the act of rising or sinking the Nautilus depended on the alternate introduction of water or air into the siphuncle, was accepted, but he was at a loss to find the source from which this air could be obtained at the bottom of the sea. The problem was not solved until some years after his death. In recognition of his researches, the species *Nautilus Parkinsoni* was named in his honour by F.E. Edwards, but by this time Parkinson had been dead for many years.

For Parkinson, Noah's Flood was now a geologically recent event which had folded the rock layers and moulded the topography. Parkinson continued his letter to show how closely this "new geology" agreed with the Book of Genesis:

A pleasing and perhaps unexpected accordance (he believes) appears between the order in which, according to the scriptural account, creation was accomplished, and the order in which the fossil remains of creation are found deposited in the superficial layers of the earth. So close indeed is this agreement, that the Mosaic account of the world is thereby confirmed in every respect except as to the age, and the distance of time between the completion of different parts of creation. These, in consequence of the literal acceptation of the word *day*, in that account, are reckoned to be much less than what every examination of the earth's structure authorizes their being supposed. If we are constrained to receive this word as descriptive of that length of time in which the planet now performs its diurnal revolution; and are to consider the words morning and evening, applied to a time when the sun is said not to have been formed, as bearing the same meaning which they now convey, it must be acknowledged that the stumbling block is immovable. But if, on the other hand, the word day be admitted as figuratively designating certain indefinite periods, in which particular parts of the great work of creation were accomplished, no difficulty will then remain. The age of the world, according to the scriptural account, will then agree with that which is manifested by the phenomena of its stratification.

I am aware that I shall obtain very little support in such a change from the critical expositors of this part of scriptural history. . . . I did not commence the inquiries contained in these volumes, without being forewarned of the great probability, that they would terminate in the establishment of certain facts, which might materially contradict the Mosaic account of the creation . . . . Unapprised of what would be the termination of this inquiry, I resolved to prosecute it with fairness; to shrink from no question, on account of its supposed tenderness; and to conceal no conclusion, however repugnant to popular opinion or prejudice. That the result would be so strongly confirmatory of the Mosaic account, I acknowledge was unexpected; and that so close an agreement should be found of the order of creation, as stated in scripture, with the actual appearance of the depth of stratification which has been examined in modern times, must satisfy or surprise every-one—Moses could not have learned this accordance from the Egyptians (Parkinson, 1804–11, Vol. III, pp. 451–52).

Organic Remains was the first systematic treatise on oryctology published in English, the first comprehensive treatise because it dealt with the organic remains of plants as well as animals, including for the first time the vertebrates. It was eagerly read by students of the sciences, the vast majority of whom were amateurs, since there were but few professionals at that period. These eager readers were not deterred by the expense of the volumes, a large sum of money that time. Up to the period 1804–1811 in England, with the exception of Woodward's "Catalogue" of the specimens now preserved in Cambridge, there was no general work on fossils. James Parkinson tried to remedy this deficiency in his work, *Organic Remains of a Former World.*

Dr. Gideon Mantell (1790–1852) wrote in 1850: "The publication of Mr. Parkinson's *Organic Remains of a Former World* at the commencement of the present century, must be regarded as a memorable event in the history of British Palaeontology; it was the first attempt to give a familiar and scientific account of the fossil relics of animals and plants, accompanied by figures of the specimens described" (Mantell, 1850, pp. 13).

Parkinson's *Organic Remains* was reprinted a number of times during and after the author's lifetime. The first volume (1804) was reprinted in 1811 with a large number of fairly minor changes to the text, and again in 1820 with a few minor changes. The second volume (1808) was also reprinted in 1820 without any changes to the text. In 1833 a complete new "second edition" was published by M.A. Nattali, again without any text changes. The book finally went out of print at the end of the 1840s, but, as already mentioned, the plates were reprinted once more as part of Mantell's *Pictorial Atlas of Fossil Remains* (Mantell, 1850). In the introduction to this work, Mantell wrote:

Although nearly forty years have since elapsed, and hundreds of geological works, of all kinds and degrees of merit, have subsequently been issued, Mr. Parkinson's plates, owing to their fidelity and beauty, are still in such request, as to induce the proprietor, Mr. Bohn, now that the work is out of print, to publish them, with the descriptions and modern names of the fossils represented.

I gladly avail myself of this opportunity to make a passing allusion to the excellent and accomplished author Mr. Parkinson. I had the pleasure and privilege of his acquaintances in my youth, immediately after the publication of the third volume. . . . He kindly showed and explained to me the principal objects in his cabinets and pointed out every source of information on fossil remains; a department of knowledge at that time but little cultivated in England, but which peculiar circumstances had contributed to render the engrossing object of my young and ardent mind. In after years Mr. Parkinson warmly encouraged my attempts to elucidate the nature of the strata and organic remains of my native county, Sussex, a district which was then supposed to be destitute of geological interest, and he revised my drawings, and favoured me with his remarks on many subjects treated of in my first work '*The Fossils of the South Downs*'

(Mantell, 1850, pp. 13–14).

Dr. Mantell sadly related the fate of Parkinson's renowned collection of fossils:

After his death (continued Dr. Mantell) his beautiful and choice collection was sold by auction. The fine series of silicified zoophytes was purchased by Mr. Featherstonhaugh, and taken to America; and some years afterwards, was destroyed by fire which consumed the museum in which it was placed (Mantell, 1850).

Parkinson's collecting instinct was not confined to fossils. He also collected minerals, coins, and medals. In his "last will and testament," he bequeathed "A cabinet of coins and medals" to his eldest son, John, "with the cases and drawers in which they are contained at my present residence or my former residence in Hoxton Square, also whatever may be due to me from Messrs. Sherwood and Jones, Booksellers, Paternoster Row, on the sale of my different publications, the Organic Remains of Former Worlds and an Introduction to the history of fossil organic remains and the engraved copper plates belonging to the above mentioned works. . . . unto my said wife, Mary Parkinson."

Among other museums to which some of Parkinson's specimens found their way ultimately were those at the Universities of Cambridge and Oxford, and a museum at Haslemere, Surrey, which had been founded by Sir Jonathan Hutchinson (1828–1912), surgeon at the London Hospital. (At the present time, however, there are none in Haslemere.)

James Parkinson was one of the 13 founding members of the Geological Society, 11 of whom met at the Freemason's Tavern, London, on November 13, 1807 for the inaugural meeting. The Society had its roots in a plan for a research institution in mineralogy and assaying, proposed by the Hon. Charles Greville in 1804 and in the British Mineralogical Society which met in London from 1799 to 1806. A more immediate impetus was the scheme to finance the publication of a monograph on calcite by the French emigré Louis de Bournon. The promoters of this scheme held regular meetings over breakfast, first at William Babington's house, and later at William Phillips'. When their immediate object was attained, it was decided to move their meetings to the evening and to hold them "at some central tavern." This led to the formation of "a little talking geological dinner club." Parkinson seems to have played no part in the earlier activities, but he was certainly present at the inaugural meeting.

Parkinson made a notable contribution to the first volume of the *Transactions of the Geological Society*, which was published in 1811 under the title "Observations on some of the Strata in the neighbourhood of London, and on the Fossil Remains contained in them" (Parkinson, 1811). It was indeed notable since it was the first work descriptive of the stratigraphical structure of the London basin, whilst at the same time it directed attention to a very important aspect of geological enquiry then coming into use, demonstrating "that certain fossils are peculiar to, and are only found lodged in particular strata," thus acknowledging and making known the labours of William Smith, the so-called "Father of English Geology".

On March 3, 1809, Parkinson was elected a member of the Committee of Maps whose function was "the construction of mineralogical maps and the collection of drawings, models, sections of mines, &c." At the annual meeting of the Society, held on February 5, 1818, he was elected a member of the Council, on which he served for two years. He appears to have been a very active member of the Society and a regular attendant at the monthly meetings, which were held at the Freemason's Tavern until, in 1810, a house at No. 8 Lincoln's Inn Fields was taken over conjointly with the Medico-Chirurgical Society of London, the former occupying the ground floor and the latter the first floor.

The description of the geology of the London basin was the best and most important of his three contributions to the *Transactions*. The majority of the papers in the *Transactions* had hitherto dealt with "structural geology, lithology or mineralogy," stated to have been "the absorbing topics of the period," but "Mr. Parkinson's contribution opened a gate which led into the new field of stratigraphical geology, a practical application of the science of oryctology."

"The study of fossil remains," he observed, "has hitherto been directed exclusively to the consideration of the specimens themselves; and hence has been considered rather as an appendix to Botany and Zoology than as (what it really is) a very important part of geological inquiry." He referred to William Smith, "who first ascertained the constancy in the order of superposition, and the continuity of the strata of this island," and added that, "Already have these examinations thus carried on, taught us the following highly instructive facts, that exactly similar fossils are found in distant parts of the same stratum, not only where it traverses this island, but where it appears again on the opposite coast; that in strata of considerable comparative depth fossils are found, which are not discovered in any of the superincumbent beds," and, he continued, "These general facts lead us to hope that geology may derive considerable assistance from an examination of the fossils, made in connection with that of the strata to which they belong" (Parkinson, 1811).

The paper was an attempt to investigate some of the upper strata in the London area together with their fossils. "Although by no means complete," wrote the author, "it will, it is hoped, induce others, who possess superior abilities and opportunities, not only to re-examine more correctly these strata, but to extend their researches to the subjacent strata" (Parkinson, 1811).

The most superficial stratum, "beds of sand and gravel," was of particular interest to Parkinson, since it was very evident in Shoreditch, Islington, and Hackney, and he described four kinds of pebbles to be found in this layer together with their organic remains. The next he termed the "blue clay" stratum, which only a year later (1812) was named by William Smith the "London clay," from the fact that it achieved its greatest thickness around London. It varied in colour from yellowish brown to dark bluish grey, and Parkinson suggested that these differences were due to the degree of oxidation of the iron present in the stratum, an observation later confirmed by geologists. "In examining this stratum," he observed, "the curious fact that certain organic remains are peculiar to particular depositions, is first observed."

His second contribution to the *Transactions of the Geological Society* was entitled: "Some Observations on the specimens of Hippurites presented to the Society by the Hon. H.G. Bennet" (Parkinson, 1814a).

He contributed several articles on geological subjects to the *British Encyclopaedia, or Dictionary of the Arts and Sciences*, edited by William Nicholson. In the preface is an announcement that "James Parkinson, Esq., author of an elaborate and extensive work on the 'Organic Remains of a Former World,' composed the articles, Geology, Oryctology, Rocks and Shells, which appear in this Dictionary" (Nicholson, 1809, Vol. I, p. 10).

His definition of the science of Geology is original and interesting. "Geology," he said, "has for its object the structure and formation of the globe: it, of course, embraces the consideration of the materials of which it is composed and the circumstances peculiar to its original formation, as well as the different states under which it has existed and the various changes it has undergone." Rocks he classified as primitive, transition, and secondary. On the agency of water in their formation, he remarks: "Indubitable evidence of the water having stood over the tops of the mountains, which are at present much above the level of the ocean, is yielded, by the circumstances of various organised beings, former inhabitants of the water, being embedded in these mountains and even in their summits." Though himself wholeheartedly on the side of the Wernerians or Diluvialists, he admits: "This mode of formation will not, however, be admitted by every geologist." Theories of the formation of the earth formulated by ancient and more recent philosophers and geologists are discussed, especially those of Werner and Hutton. Referring to the "powerful agency of fire or water in the formation of the earth," he says: "hence geologists have been rather whimsically named, according to the particular agency which they have supported in their discourses, Plutonists and Neptunists." The system of Mr. Hutton is beautifully described. "It ascribes the formation of continents, of mountains, valleys, &c., not to accidental occurrences, but to the operation of regular and uniform causes; making the decay of one part subservient to the restoration of another, by successive reproductions. Thus he supposes this globe to be regulated by a system of decay and renovation, and that these are affected by certain processes which bear a uniform relation to each other." "The expansive power of subterranean fire," he continues, "is called in to explain, by the elevation of strata, their various positions." "The system of the celebrated Werner" is contrasted with that of Hutton, and is described in detail as expounded by Werner's "learned and zealous annotator," Professor Jameson (Parkinson, 1809, Vol. III).

In the article "Rock," rock is defined as "A stony mass forming a portion of the substance of the globe." He classifies rocks according to the classification laid down by A.G. Werner (1749–1817) as Primitive, Transition, Floetz, Alluvial, and Volcanic, terms which are familiar even today. Primitive rocks were "chemical precipitates, bearing no trace of organised beings." Transition rocks were formed "during the transition of the earth into a habitable state" and contained the first traces of fossil organic remains. Floetz rocks "disposed in flat or horizontal strata, after the creation of animals and vegetables; the remains of which are often found in the substance of these rocks" (Parkinson, 1809, Vol. II).

The article on oryctology is the longest of the four articles, comprising 10,000 words, and it is accompanied by two plates, each containing eight figures of fossils. Oryctology is defined as "the science which teaches the natural history of those animal or vegetable substances which are dug out of the earth, in a mineralized state." The subject being brought down to the level of the general reader could well serve as an introduction to Parkinson's larger works on oryctology. It is interesting to note the author's continual use of the third person in his article in which he refers to "Mr. Parkinson." For example, he observes that Mr. Parkinson objects to the use of the term "fossil" which represents every mineral substance dug out of the earth (oryctology was a word used for the study of every kind of fossil); he stresses that Mr. Parkinson also objects to the term *petrification* as implying a conversion into stone only; and finally states that Mr. Parkinson considers fossils as of two kinds, primary and secondary. He discusses impartially Mr. Parkinson's theories regarding the formation of bituminous vegetable fossils by a process of fermentation and bituminization, and the lapidous vegetable fossils by impregnation with water holding earths or minerals in solution (Parkinson, 1809).

The article "*Shell*" deals with the subject of conchology, called by earlier British writers such as Dr. Martin testaceology. Shell is defined as "a substance of stony hardness, composed of carbonate of lime variously combined with animal gluten, and serving for the coverings and habitations of different animals, mostly of the order of Mullusca; allowing of the occasional protrusion of their naked body." The collection of recent shells was a fascinating and popular hobby at the beginning of the 19th century.

"The various forms, the beautiful colours and the high polish, which the shells possess," he contends, "have long rendered these objects of research to the curious naturalist: many of them possessing these properties in a high degree; and being from their variety purchased at very high prices." "Animal shells," he explains, "are naturally divided into univalves, bivalves and multivalves," and he reminds his readers that science owes much to French naturalists for a better and more extensive classification of shells.

Parkinson's opinion was much sought after upon all matters relating to fossils, and for many years he was the only palaeontologist in the Geological Society. His only rival in the field was William Martin, an honorary member of the Society, but he lived far away at Macclesfield in Derbyshire. Of the 13 original members of the Society, it is said that "Parkinson was the representative of palaeontology, William Phillips of Stratigraphical Geology, and Greenough. . . . of General Geology."

In 1822, Parkinson published *Outlines of Oryctology. An Introduction to the Study of Fossil Organic Remains; especially of those found in the British Strata: intended to aid the student in his inquiries respecting the nature of Fossils, and their connection with the formation of the Earth* (Parkinson, 1822).

He had contemplated publishing this work much earlier, according to the following announcement which appeared in *Nicholson's Journal* in 1810 (Vol. 25, p. 317).

> Mr. Parkinson has withdrawn his *Introduction to the Knowledge of Fossils*, announced at the end of the first volume of *Organic Remains of a former World*, considering its publication as entirely superseded by Mr. Martin's excellent *Outlines* on the same subject. The third volume of *Organic Remains* is in considerable forwardness.

This little octavo volume containing 343 pages, together with nine plates crammed with 220 figures of fossils drawn and engraved by Springsguth, it was probably a larger, more comprehensive, and more up-to-date version of the work which might have been published 12 years earlier.

William Martin, an excellent naturalist, a former actor who eked out a living as a drawing master, died by his own hand in the year 1810. It is probable that a philanthropic motive stood behind Parkinson's suspension of the publication of *Outlines of Oryctology*, as poor William Martin had left a large family totally unprovided for.

In the preface to *Outlines of Oryctology*, Parkinson says

> For the several imperfections in this work, which may have escaped the author's attention, he craves indulgence; hoping that they will not be found of such importance as to render the wish too presumptious of having it considered as a humble subsidiary to that scientific and most reliable comprehensive work *Outlines of the Geology of England and Wales*, by the Rev. W.D. Conybeare and W. Phillips.

This classic, the first geological textbook descriptive of the stratigraphy of England and Wales, was published but a few months before Parkinson's *Outlines of Oryctology*, in 1822. Unfortunately the work was never completed by a second volume as intended. Parkinson had designed his book as a kind of practical field manual for the student of palaentology, for he said he hoped it would "become a useful *vade mecum* for the intelligent traveller." As compared with his previous definitions of the term oryctology to be found in *Organic Remains* or Nicholson's encyclopaedia, it is now much wider and more explicit. It is "the science which enquires into the nature, origin, and formation of those bodies which possess the figures, markings or structure of vegetables or animals, whilst their substance evinces their having been preserved through many ages, by certain changes affected in subsequence or subterranean situations" (Parkinson, 1822, p. 7).

Only 33 pages are devoted to the classification and description of vegetable fossils. It is notable that he has not changed his speculative opinions regarding the formation of peat and Bovey coal as set down in *Organic Remains,* but adds that "impartiality here requires that the opinion of Professor Jameson on this subject should be noticed." He now has the advantage of a recent full classification of the genera and species of Zoophytes advanced by the French naturalist, Lamouroux (1816), and a new classification of the encrinites and pentacrinites in accordance with the researches of J.S. Miller of Bristol, *A Natural History of the Crinoidea* (1821), concerning which he states:

Whilst this part of the present work was in the hands of the printer I had the gratification of receiving Mr. Miller's *Natural History of the Crinoidea*, or Lily-shaped Animals: the pleasure which I experienced in contemplating the scientific and successful inquiries of this subject on which, as a student, I had myself toiled, led me to hope that by pointing out the importance and extent of his discoveries, I should diffuse pleasure among my readers, and by giving them a glimpse of the treat there prepared for them, render them eager to partake of a banquet so rich and so cheap, it containing 50 illustrated lithographic plates (J. Parkinson, Esq. MGS, Hoxton Square, London was among subscribers listed in the book.) (Parkinson, 1822, p. 8; 94).

After Parkinson's death, a second edition (with the author's latest corrections) was issued in 1830 by a firm of publishers to whom the rights had been sold, and a third edition appeared a few years later. It is believed that this book, his last, sold in larger numbers than any of his previous publications. A reviewer said that the book was "to the learner an easy and complete introductory manual: and to the well informed a text book of convenient reference."

'One of the first steps in the chronological reform of tertiary formations' wrote Charles Lyell (1797–1875) 'was made in 1811 by an English naturalist, Mr. Parkinson, who pointed out the fact that certain shelly strata, previously termed 'Crag' in Suffolk, were decidedly imposed on a deposit which was the continuation of the blue clay of London. At the same time he remarked that the fossil testacea in these newer beds were distinct from those of the underlying blue clay, and that while some of them were of unknown species, many others were identical with species now inhabiting the British Seas.'

It is notable that nine of the thirteen original members of the Geological Society became Fellows of the Royal Society, and it might be contended that Parkinson highly deserved the honour of election to the Society in view of his researches and publications on oryctology and geology. It is true that he never contributed to the *Philosophical Transactions*, but neither had the vast majority of the Fellows. Two of his friends and neighbors in Hoxton Square were Fellows who could have nominated him, as well as his friends in the Geological Society or the medical profession, who were also Fellows. Could it have been his reticence that precluded him from putting his name forward? It might well have also been the result of a disagreement which had arisen between the Royal and Geological Societies. Sir Joseph Banks (1743–1820), President of the Royal Society, who was a member of the Geological Society, resigned in 1809 because, in his opinion, the Geological Society "had deviated from the principles which they had entertained on the first establishment." Banks desired the Geological Society to be a subsidiary society, subservient to the older Royal Society, and a plan had been drawn up for incorporating it as an "assistant society." A motion was put forward, and it was resolved that: "any proposition to render the society dependent upon or subservient to any other society does not correspond with the conception this meeting entertains of the original principles upon which the Geological Society was founded."

The controversy split the Society into those who voted for the motion, and those against. In consequence of the decision against amalgamation, Sir Humphry Davy, a Fellow of the Royal Society, also resigned though he returned to the Society later; Sir Joseph Banks, however, never returned. Parkinson was a participant in the conflict and voted for the motion against amalgamation, and how right he was in this decision, time has shown. Sir Roderick Impey Murchison (1792–1871), President of the Geological Society, referred in his presidential address in 1842 to the correctness of the decision of the Society to remain independent in these words: "The advantage which science has reaped through the independence of action and division of labour is now, indeed, admitted even by those who were opponents and have lived to see our success."

Although more than 10 years had elapsed since Parkinson was associated with the unpopular movement for parliamentary reform, could this have been the reason why he was not admitted to the august Royal Society? His scientific attainments fully merited recognition at a time when the vast majority of its 600 or 700 Fellows were described as "useless" and having no scientific pretensions.

If the Royal Society did not recognize Parkinson's achievements in the science of Palaeontology, his fellow surgeons took note of his surgical achievements. In 1823 the Royal College of Surgeons awarded him the

Honorary Gold Medal of the College, and although that medal had been instituted in 1802, Parkinson was the first recipient of this honour. An extract from the minutes of the court of Assistants of the College states

> Eleventh day of October 1822. In pursuance of the Recommendation from the Board of Curators, contained in the minutes of the Board now read, of granting to Mr. Jas. Parkinson, of Hoxton Square, the Honorary Medal of the College in consideration of his useful labours for the promotion of Natural Knowledge, particularly expressed by his splendid work on Organic Remains and of his liberal and valuable information when called upon by the College in its research for Facts relating to the scientific designs.

A nomination and ballot followed and the minutes of December 13 record that: "The Honorary Medal of the College was granted to James Parkinson, Esquire, of Hoxton Square." The presentation ceremony took place on the 11th of April 1823, his 68th birthday. He had the honour of dining with members of the Council beforehand, a resolution having been passed:

> That the Regulation of this Council prohibiting the Invitation of any Member of the College in actual practice to dine with Members of the Council be suspended in the person of Mr. Parkinson on this occasion.

At the time of the presentation he had moved from No. 1 Hoxton Square and was then living at No. 3 Pleasant Row in the Kingsland Road. He was still the Parish Doctor, still actively pursuing the study of fossils. He had yet one more medical article to write before his death the following year.

CHAPTER 12

A Discussion of Parkinson's *Essay on the Shaking Palsy*

James Parkinson's tract: *An Essay on the Shaking Palsy* (Parkinson, 1817), has deservedly become a medical classic. Copies of the original edition of 1817 are now rare; but of all his writings it is this short work that has endured, both for the felicity of its style and for the accuracy with which the clinical history of the disease was described. The book was well received at the time of its publication, and testaments to Parkinson's diagnostic acumen recurred with increasing frequency during the course of the 19th century, as neurology and psychiatry became established as professional disciplines.

Thus Thomas Buzzard (1831–1919), in his *Clinical Lectures on Diseases of the Nervous System* (Buzzard, 1882), given in the early 1880s at the Hospital for the Paralysed and Epileptic, London, employed Parkinson's definition and description of paralysis agitans in his presentation of cases, stating that

> The disease "shaking palsy", or "paralysis agitans" . . . was first regularly described by our countryman Parkinson in 1817. Parkinson was a member of the Royal College of Surgeons, and his *Essay on the Shaking Palsy* presents so graphic and admirable a description of the disease that comparatively little has been left for subsequent observers to add to his account (Buzzard, 1882, pp. 329).

Despite this encomium, it is notable that Buzzard did not refer to "Parkinson's disease" as such, although the eponymous term, "la maladie de Parkinson," coined by Charcot, had been common usage amongst Charcot's colleagues and pupils at the Salpêtrière for some two decades. Buzzard's reluctance to adopt the eponym was characteristic of British clinicians of the time. William Gowers (1845–1919) was to put on record his preference for the retention of the name "shaking palsy" in the summa of 19th century neurology, his *Manual of the Diseases of the Nervous System* (Gowers, 1886–88).

The history of the gradual recognition of Parkinson's description of paralysis agitans as indeed constituting a distinct disease that was diagnostically differentiable from the forms of tremor with which it had hitherto been confounded is an interesting one, and will be discussed in detail below. Parkinson's achievement, and the originality of the *Essay on the Shaking Palsy*, must be judged in comparison with the earlier medical literature of chorea, scelotyrbe, tremor, and palsy. In the second and third chapters of the *Essay*, Parkinson carefully assessed the theories of tremor which had been provided by 17th and 18th century medical authorities Sylvius, Juncker, Van Swieten, and Gaubius. He analysed the presentation of the symptoms characteristic of paralysis agitans in the nosological writings of Sauvages, Linneaus, Vogel, Cullen, and MacBride. A conspectus of this literature, following Parkinson's concise account, is given in the outline of the *Essay* that follows; salient extracts from the *Essay*, those passages that most frequently attracted the attention of subsequent neurologists, are liberally quoted.

An Essay on the Shaking Palsy was published as an octavo volume of 66 pages of text, divided into five chapters, with a four page preface. The chapters bear the following summary headings:

Chap. I. Definition—History—Illustrative cases.
Chap. II. Pathognomonic symptoms examined—Tremor Coactus—Scelotyrbe Festinians.
Chap. III. Shaking Palsy distinguished from other diseases with which it may be confounded.
Chap. IV. Proximate cause—Remote causes—Illustrative cases.
Chap. V. Considerations respecting the means of cure.

In the preface, Parkinson tendered "some conciliatory explanation" for publishing the *Essay*, in the hope "that the offering of the following pages to the attention of the medical public, will not be severely censured." Shaking Palsy, he emphasises,

> is of a nature highly afflictive. Notwithstanding which, it has not yet obtained a place in the classification of nosologists; some have regarded its characteristic symptoms as distinct and different diseases, and others have given its name to diseases differing essentially from it; whilst the unhappy sufferer has considered it as an evil, from the domination of which he had no prospect of escape (Parkinson, 1817, p. ii).

As the disease is of long duration, continued observation of the same case over several years was required in order to establish a correct history of the symptoms of Shaking Palsy at each stage of its course. Parkinson had the opportunity of so doing: here, the first case history of the six provided in Chapter I, that of a gardener aged 50 years, is alluded to. From these particular details Parkinson abstracted his classic generalised description of paralysis agitans. From these repeated observations of cases in which the disease existed in different stages of its progress, Parkinson hoped that he "had been led to a probable conjecture as to the nature of the malady, and that analogy had suggested such means as might be productive of relief, and perhaps even of cure, if employed before the disease had been too long established" (Parkinson, 1817, p. iii). This refers to the final two chapters of the *Essay*, in which Parkinson tentatively identified as the proximate cause of the disease a lesion of the cervical portion of the spinal cord, extending to the medulla oblongata. He suggested a treatment involving blistering and the withdrawal of blood from the upper part of the neck. In the rationale for his diagnosis and therapy, Parkinson acknowledges "that mere conjecture takes the place of experiment; and that analogy the substitute for anatomical examination, the only sure foundation for pathological knowledge" (Parkinson, 1817, p. i). With this prefacing caveat, the *Essay* was addressed to the "friends to humanity and medical science, who have already unveiled . . . many of the morbid processes by which health and life is abridged" in order to prompt their further researches.

So slight and nearly imperceptible are the first inroads of this malady, and so extremely slow is its progress, that it rarely happens, that the patient can form any recollection of the precise period of its commencement. The first symptoms perceived are, a slight sense of weakness, with a proneness to trembling in some particular part; sometimes in the head, but most commonly in one of the hands and arms. These symptoms gradually increase in the part first affected; and at an uncertain period, but seldom in less than twelve months or more, the morbid influence is felt in some other part. Thus assuming that one of the hands and arms to be first attacked, the other, at this period becomes similarly affected. After a few more months the patient is found to be less strict than usual in preserving an upright posture; this being most observable whilst walking, but sometimes whilst sitting or standing. Sometime after the appearance of this symptom, and during its slow increase, one of the legs is discovered slightly to tremble, and is also found to suffer fatigue sooner than the leg on the other side. And in a few months this limb becomes agitated by similar tremblings, and suffers a similar loss of power.

Hitherto the patient will have experienced but little inconvenience, and befriended by the strong influence of habitual endurance, would perhaps seldom think of his being the subject of disease, except when reminded of it by the unsteadiness of his hand, whilst writing or employing himself in any nicer kind of manipulation. But as the disease proceeds, similar employments are accomplished with considerable difficulty, the hand failing to answer with exactness the dictates of the will. Walking becomes a task which cannot be performed without considerable attention. The legs are not raised to that height, or with that promptitude which the will directs, so that the utmost care is necessary to prevent frequent falls.

At this period the patient experiences much inconvenience, which unhappily is found daily to increase. The submission of the limbs to the directions of the will can hardly ever be obtained in the performance of the most ordinary offices of life. The fingers cannot be disposed of in the proposed directions, and applied with certainty to any proposed point. As time and the disease proceed, difficulties increase: writing can now hardly be at all accomplished; and reading, from the tremulous motion, in accomplished with some difficulty. Whilst at meals the fork not being duly directed frequently fails to raise the morsel from the plate:

which, when seized, is with much difficulty conveyed to the mouth. At this period the patient seldom experiences suspension of the agitation of the limbs. Commencing, for instance in one arm, the wearisome agitation is borne until beyond sufference, when by suddenly changing the posture it is for a time stopped in that limb, to commence, generally, in less than a minute in one of the legs, or in the arm of the other side. Harrassed by this tormenting round, the patient has recourse to walking, a mode of exercise to which the sufferers from this malady are in general partial; owing to their attention being thereby somewhat diverted from their unpleasant feelings by the care and exertion required to ensure its safe performance.

But as the malady proceeds, even this temporary mitigation of suffering from the agitation of the limbs is denied. The propensity to lean forwards becomes invincible, and the patient is thereby forced to step on the toes and fore part of the feet, whilst the upper part of the body is thrown so far forward as to render it difficult to avoid falling on the face. In some cases, when this state of the malady is attained, the patient can no longer exercise himself by walking in his usual manner, but is thrown on the toes and fore part of the feet; being, at the same time, irresistibly impelled to take much quicker and shorter steps, and thereby to adopt unwillingly a running pace. In some cases it is found necessary entirely to substitute running for walking: since otherwise the patient, on proceeding only a very few paces, would inevitably fall.

In this stage, the sleep becomes much disturbed. The tremulous motion of the limbs occur during sleep, and augment until they awaken the patient, and frequently with much agitation and alarm. The power of conveying the food to the mouth is at length so much impeded that he is obliged to consent to be fed by others. The bowels, which had been all along torpid, now, in most cases, demand stimulating medicines of very considerable power: the expulsion of the faeces from the rectum sometimes requiring mechanical aid. As the disease proceeds towards its last stage, the trunk is almost permanently bowed, the muscular power is more decidedly diminished, and the tremulous agitation becomes violent. The patient walks now with great difficulty, and unable any longer to support himself with his stick, he dares not venture on this exercise, unless assisted by an attendant, who walking backwards before him, prevents his falling forwards, by the pressure of his hands against the forepart of his shoulders. His words are scarcely intelligible; and he is no longer able to feed himself, but when the food is conveyed to the mouth, so much are the actions of the muscles of the tongue, pharynx, etc. impeded by impaired action and perpetual agitation, that the food is with difficulty retained until masticated; and then as difficulty swallowed. Now also, from the same cause, another unpleasant circumstance occurs: the saliva fails of being directed to the back part of the fauces, and hence is continually draining from the mouth, mixed with the particles of food, which he is no longer able to clear from the inside of the mouth.

As the debility increases and the influence of the will over the muscles fades away, the tremulous agitation becomes more vehement. It now seldom leaves him for a moment; but even when exhausted nature seizes a small portion of sleep, the motion becomes so violent as not only to shake the bedhangings, but even the floor and sashes of the room. The chin is now almost immoveably bent down upon the sternum. The slops with which he is attempted to be fed, with the saliva, are constantly trickling from the mouth. The power of articulation is lost. The urine and faeces are passed involuntarily; and at the last, constant sleepiness, with slight delirium, and other marks of extreme exhaustion, announced the wished-for release (Parkinson, 1817, pp. 3–9).

The remainder of Chapter I is taken up with accounts of six cases of Shaking Palsy known to Parkinson. The last was later referred to by other doctors because of the unusual course taken by the disease.

In a case which presented itself to observation since those above-mentioned, every information as to the progress of the malady was very readily obtained. The gentleman who is the subject of it is seventy-two years of age. . . . About eleven or twelve or perhaps more, years ago, he first perceived weakness in the left hand and arm, and soon after found the trembling commence. Years afterwards the right arm became affected in a similar manner: and soon afterwards the convulsive motions affected the whole body, and began to interrupt the speech. In about three years from that time the legs became affected. Of late years the action of the bowels had been very much retarded; and at two or three different periods had, with great difficulty, been made to yield to the action of very strong cathartics. But within the last twelve months this difficulty has not been so great; perhaps owing to an increased secretion of mucus, which envelopes the passing faeces, and which precedes and follows their discharge in considerable quantity.

About a year since, on waking in the night, he found that he had nearly lost the use of his right side, and that the face was much drawn to the left side. His medical attendant saw him the following day, when he found him languid, with a small and quick pulse, and without pain in the head or disposition to sleep. Nothing more therefore was done than to promote the action of the bowels, and apply a blister to the back of the neck, and in about a fortnight the limbs had entirely recovered from their palsied state. During the time of their having remained in this state, neither the arm nor the leg of the paralytic side was in the least affected with the tremulous agitation; but as their paralysed state was removed, the shaking returned.

At present he is almost constantly troubled with the agitation, which he describes as generally commencing in a slight degree, and gradually increasing, until it arises to such a height as to shake the room; when, by a sudden and somewhat violent change of posture, he is almost always able to stop it. . . . When the agitation had not been thus interrupted, he stated, that it gradually extended through all the limbs, and at last affected the whole trunk . . . (Parkinson, 1817, pp. 14–17).

In Chapter II, Parkinson remarks that "certain affections, the tremulous agitations, and the almost invincible propensity to run, when wishing only to walk, each of which has been considered by nosologists as distinct diseases, appear to be pathognomonic symptoms of this malady" (Parkinson, 1817, p. 19). He goes on to outline some of the ideas held by his predecessors, including F. Sylvius de la Boe (1614–1667), Johann Juncker (1680–1759), G. van Swieten (1700–1772), and F.B. de Sauvages (1706–1767). He also outlined some earlier ideas concerning the other major symptom of the Shaking Palsy; the "Propensity to bend the trunk forwards, and to pass from a walking to a running pace" (Parkinson, 1817, p. 24).

In Chapter III, Parkinson presents the evidence for the Shaking Palsy being distinct nosologically from other diseases involving tremor. Of considerable importance to him is the circumstance that in other diseases.

If the trembling limb be supported, and none of its muscles be called into action, the trembling will cease. In the real Shaking Palsy the reverse of this takes place, the agitation continues in full force whilst the limb is at rest and unemployed; and even if sometimes diminished by calling the muscles into employment (Parkinson, 1817, p. 32).

Tremors which are liable to be confused with the real Shaking Palsy include those resultant from alcohol abuse, tea and coffee abuse, and old age. Parkinson quotes at length from Thomas Kirkland's (1722–1792) *Commentary on Apoplectic and Paralytic Affections* (Kirkland, 1792).

Chapter IV considers the proximate and remote causes of the disease. Parkinson attributes the Palsy to

A diseased state of the *medulla spinalis*, in that part which is contained in the canal, formed by the superior cervical vertebrae, and extending, as the disease proceeds, to the *medulla oblongata* (Parkinson, 1817, pp. 33–34).

However, Parkinson is careful to point out that he has never performed a postmortem on a case and that

Conjecture founded on analogy, and an attentive consideration of the peculiar symptoms of the disease, have been the only guides that could be obtained for this research, the result of which is, as it ought to be, offered with hesitation (Parkinson, 1817, p. 33).

He argues that because the disease affects so many parts "the injury is rather in the source of this influence than merely in the nerves of the parts," and that since the senses and the intellect remain unimpaired "the morbid state does not extend to the encephalon" (Parkinson, 1817, p. 34).

He quotes at some length the case of Count de Lordat, as reported by Matthew Maty (1718–1776) in the *Medical Observations and Inquiries* (1757–1784, vol. 3). The patient suffered neck and head injuries as a result of a coach crash, and in later years exhibited symptoms approximating the Shaking Palsy. After his death an autopsy was carried out. For many years this examination was falsely attributed to Parkinson and remained the only such examination of Shaking Palsy. Parkinson concedes that de Lordat may not have been suffering from Shaking Palsy but suggests that his case "serves at least to show that nearly the same parts were the seat of the disease in both instances. Thus we attain something like confirmation of the supposed proximated cause, and of one of the assumed occasional causes" (Parkinson, 1817, p. 45).

Parkinson next considers what part rheumatism might play in bringing on the Shaking Palsy. Two of the cases he has described in Chapter I suffered from "rheumatic affections" and he is aware of others. He describes the case of "A.B." in detail. This patient had suffered "rheumatic affection of the deltoid muscle" which led to the appearance of symptoms resembling the early stages of the Shaking Palsy. Eventually, however, these cleared up without further development. Parkinson concludes

> The commencement, progress, and termination of this attack; with the success attending the mode of treatment, and the symptoms which followed, seem to lead to the conjecture, that the proximate cause of the disease, in this case, existed in the medulla spinalis, and that it might, if neglected have gradually resolved itself into that disease which is the object of our present inquiry (Parkinson, 1817, p. 50).

In the concluding chapter Parkinson ponders "the means of cure." Characteristically, he is optimistic:

> . . . there appears to be sufficient reason for hoping that some remedial process may ere long be discovered, by which, at least, the progress of the disease may be stopped (Parkinson, 1918, pp. 56–57).

He expresses his highest hopes for those treated early but feels that even in the later stages something might be gained by suitable treatment.

The treatment he recommends is as follows:

> . . . blood should first be taken from the upper part of the neck, unless contra-indicated by any particular circumstance. After which vesicatories should be applied to the same part, and a purulent discharge obtained by appropriate use of the Sabine Liniment; having recourse to the application of a fresh blister, when from the diminution of the discharging surface, pus is not secreted in a sufficient quantity (Parkinson, 1817, pp. 58–59).

Parkinson ends the book with a justification of the practise of dissecting corpses in order to gain knowledge of disease processes. It is a passage much quoted in the period after its publication, and is worth reprinting in full:

> Before concluding these pages, it may be proper to observe once more, that an important object proposed to be obtained by them is, the leading of the attention of those who humanely employ anatomical examination in detecting the causes and nature of diseases, particularly to this malady. By their benevolent labours its real nature may be ascertained, and appropriate modes of relief, or even of cure, pointed out. To such researches the healing art is already much indebted for the enlargement of its powers of lessening the evils of suffering humanity. Little is the public aware of the obligations it owes to those who, led by professional ardour, and the dictates of duty, have devoted themselves to these pursuits, under circumstances most unpleasant and forbidding. Every person of consideration and feeling, may judge of the advantages yielded by the philanthropic exertions of a HOWARD; but how few can estimate the benefits bestowed on mankind, by the labours of a MORGAGNI, HUNTER, or BAILLIE (Parkinson, 1817, pp. 65–66).

The business of dissecting corpses was controversial, especially since much of the raw material of these researches was supplied through the questionable and much despised activity of the so-called "Resurrection Men." While avoiding specific references, Parkinson seems here to be hinting that since medical science cannot progress without such research, the supply of corpses must be better organized and controlled. In this sense he may be said to be looking forward to the Anatomy Act of 1832, which finally brought the whole matter under proper regulation.

The *Essay* received very favourable reviews in the medical press following its publication. *The Medico-Chirurgical Journal and Review* wrote, "Mr. Parkinson has had the opportunity of observing the range of symptoms in this tedious disease for several years, and has hence been led particularly to observe several other cases in which the disease existed in different stages of its progress. By these repeated observations, he thinks he has been led to a probable conjecture as to the true nature and proper mode of treatment, and justly considers it his duty to give publicity to his opinions. Mr. Parkinson needs to be under no dread of any censure which the precipitated publication of mere conjectural suggestions may incur; his pamphlet does not come under any such title; and the name of the author would be a sufficient passport to publicity, and security from aspersion, for a much less reputable performance." The review occupied several pages and gave an excellent summary of the pamphlet chapter by chapter, ending with the remark that, "The reader will now perceive, that this little pamphlet, is highly worthy of perusal, and deserves the attention of the medical public" (1817, Vol. IV pp. 401–08).

The Monthly Gazette of Health, under the heading Shaking Palsy, said that, "Surgeon Parkinson, a practitioner of considerable experience and scientific attainments has lately published a popular treatise on this disease. After investigating the causes, etc., of this malady, all he has ventured to assume is, that the disease depends on a disordered state of that part of the spinal column which is contained in the spine of the neck; but of what nature that morbid change is, and whether originating in the spinal marrow itself, in its membranes, or in the containing canal, remains a subject of doubt and conjecture. But although at present, uninformed as to the precise nature of the disease, still it ought not to be considered as one against which there is no countervailing remedy."

The hypothesis as to the underlying pathology of the disease seems to have been more particularly noticed by the reviewers rather than Parkinson's description of the signs and symptoms.

Several years elapsed before the teachers of medicine in the London medical schools began to take notice of Parkinson's identification of the Shaking Palsy as a clinical entity. An early reference to his work was made by Dr. John Elliotson (1791–1868) in a clinical lecture delivered at St. Thomas' Hospital on October 11, 1830, six years after Parkinson's death. Elliotson said, "The best account of this disease which I have seen is one given by a general practitioner, now deceased, of the name of Parkinson, a highly respectable man, who wrote an essay upon the subject in 1817, from which I have derived nearly all I know upon the complaint. . . . The words of Mr. Parkinson in describing this disease are, "Involuntary tremulous motion, with lessened muscular power, in parts not in action and even when supported; with a propensity to bend the trunk forwards, and to pass from a walking to a running pace: the senses and intellect being uninjured. . . . It does not appear as if this disease of which I am at present speaking, was well characterised or distinguished before Mr. Parkinson wrote on the subject. The peculiar characters which mark the shaking of drunkards, as different from the disease of muscular agitation, had been pointed out by Galen, and many others since his time, but paralysis agitans was not well defined before the Essay of Mr. Parkinson" (Elliotson, 1830, p. 120).

Paralysis agitans had been described as a distinctive disease in *Elements of the Theory and Practice of Physic* (Gregory, 1828). A footnote to the section on this disease is believed to be the first reference to Parkinson's pamphlet, "vide Parkinson's Essay on the Shaking Palsy".

Dr. Marshall Hall (1790–1857), in a lecture on paralysis agitans given at Guy's Hospital in 1837, said, "I must now draw your attention very briefly, to another disease of the spinal marrow, the paralysis agitans. Its symptoms have been well described by Mr. Parkinson, but its morbid anatomy has not been traced" (Hall, 1838, p. 41).

Dr. Thomas Watson (later Sir Thomas Watson, Bart, 1792–1882), lecturing at King's College, London, in 1841 on paralysis agitans, told his auditors, "I refer to what has been called the *shaking palsy—paralysis agitans*. Allusions to this form of disease are to be found in many of the older systematic writers on physic; but it never was much attended to in this country until Mr. Parkinson published an essay upon it in the year 1817; and a very interesting little pamphlet it is." After giving Parkinson's definition of the disease, Watson proceeded: "Mr. Parkinson conjectures that this complaint results from some chronic change of the upper part of the spinal cord, or the medulla oblongata: but dissections are wanting to support or refute that conjecture . . . a more exact pathology of the shaking palsy is still needed. Dr. Hall observes that the symptoms have in several particulars, a marked resemblance to the effects observed by M. Serres (and related by him in his *Anatomie Compareé du Cerveau*)* of disease of the tuber annulare, or of the tubercula quadrigemina." 24 years had elapsed since the appearance of the pamphlet and no account of another autopsy was yet forthcoming. Marshall Hall had given the first hint that the seat of the disease might be in the brain and not the spinal cord. Dr. Watson continued, "Nor have we any ascertained means of curing the disease. Dr. Elliotson indeed says that he has succeeded in one instance (of which,

*E.R.A. Serres, *Anatomie Comparée du Cerveau dans les Quatre Classes des Animaux Vertebrés* (1824, p. 26).

however, the particulars are not given) with the carbonate of iron: but that he had tried the same medicine in vain in several other cases" (Watson, 1843, p. 662–64).

David Maclachlan (1807–1870) in his book, *Diseases and Infirmities of Advanced Life* (Maclachlan, 1863), refers to "Mr. Parkinson whose interesting essay must ever be referred to, as giving a faithful account of the symptoms of the disease from beginning to the end, and is still the best work on the subject" (Maclachlan, 1863, p. 21). Maclachlan had under his care in the Chelsea Royal Hospital a case of paralysis agitans in an old army pensioner who had developed the disease at the age of 60 and who lived to the age of 107.

In the middle of the 19th century the study of paralysis agitans attracted the attention of French physicians. Dr. Armand Trousseau (1801–1867), physician to the Hôtel Dieu at Paris, differentiated paralysis agitans from chorea. He observed that, "there is no paralysis at the commencement of this strange form of chorea, which is so inappropriately named paralysis agitans" (Trousseau, 1868, p. 444). In a lecture on "Senile Trembling and Paralysis Agitans," he said of the latter, "paralysis agitans which some authors of eminence have confounded, not without some reason with chorea festinans . . ." (Trousseau, 1868, p. 445). He lamented the fact that anatomical lesions peculiar to paralysis agitans had never been studied in France, and remarked, "We must pay great attention to the alterations which Parkinson, Oppolzer and Lebert have described" (Trousseau, 1868, p. 446).

In 1860 Professor Johann Oppolzer (1808–1871) of Vienna had performed a post mortem on a case of paralysis agitans, the first to be recorded after the one attributed to Parkinson. Oppolzer's case was that of a man of 72 who was afflicted with paralysis agitans following shock, produced by the bombardment of Vienna in 1848. The autopsy showed, "The pons varolii and the medulla oblongata were very manifestly indurated. Microscopical examination:—there was found in the substance of the pons varolii and of the medulla oblongata an abnormal production of connective tissue accounting for the induration of the parts. The opaque striae in the lateral columns of the cord were due to the presence of connective tissue in process of development." With the aid of microscopy, it was now possible to explain the hardening of the cord and medulla, and Trousseau commented of Oppolzer's findings that "All these changes, as well as analogous ones noted by Parkinson and Lebert, were the result of an hypertrophy of the connective tissue which enters into the composition of the nervous tissue. The hyperformation had produced compression of the nervous elements whence their atrophy and fatty degeneration. Such alterations, attended with induration of the affected parts, are termed sclerosis. In the cases in which dissection has shown incipient softening of the columns of the cord, in the same regions, this may perhaps have been a consequence merely of hyperaemic and vascular dilatation, which cause great modification in the nutrition of nervous elements. Such alterations account for the powerlessness of treatment, for none as yet seems to have been attended with certain and continued success" (quoted in Trousseau, 1868, pp. 446–49).

Trousseau also drew attention to rigidity as a phenomenon apparent in paralysis agitans. If but scant attention had hitherto been given to the pathology of the disease, there was no dearth of remedies which had been given a trial, and Trousseau summarised them at the conclusion of his lecture. Elliotson had tried carbonate of iron, others sulphur baths and potassium iodide. He himself had used spirits of turpentine, combined with hydropathy, but without curing a single case. Dr. Ramskill claimed to have affected a cure by administering cod liver oil and the phosphorised oil of the Russian pharmacopaeia. Dr. Russell Reynolds had brought about a complete cessation of the jactitations by electricity given through a Pulvermacher's chain. Dr. Handfield Jones, in his book *Clinical Observations on Functional Nervous Diseases* (Jones, 1864, p. 267), had described the cure of a case by administering Tincture of Hyoscyamus. Trousseau, however, scathingly condemned the criteria of cure established by all the writers who claimed to have effected cures.

Jean-Martin Charcot (1825–1893) left a masterly account of paralysis agitans. In a lecture given at the Hôpital Salpêtrière in Paris, he revealed that he was the first to differentiate disseminated sclerosis from paralysis agitans. He described a group of cases in which "tremor is only shown when an intentive movement is made," and contrasted it with another group in which, "tremor is a constant symptom, from whom it rarely departs except during sleep." He said that Galen had distinguished two kinds of trembling, one of which was designated tremor, the paralytic shake, and the other palpitation, a clonic, spasmodic convulsive action. Van Swieten also recognised the difference between the tremor that persists during repose in bed, a convulsive phenomenon which he called tremor coactus, and the tremor which is shown only during the execution of voluntary movements, which he defined as a paralytic tremor, tremor debilitate. Parkinson in his essay had quoted these same two observers.

Of paralysis agitans, Charcot remarked, "Its history, however, does not reach far back. The first regular description of it only dates from 1817; it is due to Dr. Parkinson, who published it in a little work entitled *Essay on the Shaking Palsy*. From that period, paralysis has often been mentioned in England and Germany; but in France it was almost unheard of until these latter years, for if I mistake not, it was first described in an explicit manner by M.G. See in his memoir on Chorea, where it figures among the diseases which might be confounded with St. Vitus' Dance. In 1859, M. Trousseau in his lecture on Chorea succinctly tabulated the principal characters of paralysis agitans. Three years later M. Vulpian and I published a study on the subject in the *Gazette Hebdomadaire* . . . We were struck by the insufficiency of the details to be found in existing works . . . and, uniting these with the accounts given by foreign authors, we traced a tolerably complete history of paralysis agitans, considering the period. From that date this disease obtained the right of domicile in classic works." He referred to articles on paralysis agitans in the second edition of Trousseau's lectures, in which considerable space was given to its consideration, the last edition of M. Grisolle's book, and in J. Russell Reynolds' *System of Medicine* (article by W.R. Sanders Vol. II, p. 184). The latter was the first complete and extensive consideration of paralysis agitans to be found in any English textbook (Charcot, 1877, pp. 129–156).

". . . but in all these descriptions, and our own does not at all escape this reproach, there is complete confusion between paralysis agitans and disseminated sclerosis. The line of demarcation between the two diseases was for the first time indicated by myself, if I mistake not, in the thesis of M. Ordenstein "(*Sur la paralysie agitante et la sclérose en plaques généralisée.* Thèse de Paris, 1868).

It was Charcot who described so aptly a peculiar characteristic of the tremor of the hands where the thumb moves over the fingers, "as when a pencil or paper ball is rolled between them: . . . [or] in crumbling a piece of bread" (Charcot, 1877, p. 138).

The characteristic peculiarity of the handwriting in patients suffering from paralysis agitans was first demonstrated by Gubler, who described the irregular and sinous strokes, the down strokes firm and nearly normal, the up strokes all tremulous.

Charcot believed that he had also contributed to the clinical picture of this disease in recognising the presence of rigidity. "We shall now point out a characteristic," he said, "which, we believe, was overlooked by Parkinson as well as by most of his successors: we allude to the *rigidity* to be found, at a certain stage of the disease, in the muscles of the extremities of the body, and, for the most part, in those of the neck also. When this symptom declares itself, the patients complain of cramps, followed by stiffness, which, at first transient is afterwards more or less lasting, and is subject to exacerbations. Generally, the flexor muscles are the first, as they are always the most intensely, affected. When the muscular stiffness has become permanent it causes the patients, in many cases, to assume a characteristic attitude. Thus on account of the rigidity of the anterior muscles of the neck, the head, as Parkinson remarked, is greatly bent forward, and, as one might say, fixed in that position: for the patient cannot, without much effort, raise it up, or turn it to the right or left. The body is also slightly inclined forwards, when the patient is standing" (Charcot, 1877, p. 129).

In a note to the second edition of the lectures, Charcot employed the eponymous name for paralysis agitans, when, in describing an illustrative case, he said, "A man aged fifty years was attacked by PARKINSON'S DISEASE." This case was of special interest because it was diagnosed as paralysis agitans in the absence of tremor, all the symptoms and signs being present, including rigidity. In this connection he also mentioned that Mr. Gowers, of the National Hospital, London, had a similar case, a patient of the name of Anne Phillips, in which all the characteristics were present except trembling, which was hardly perceptible.

"Paralysis Agitans," said Charcot, "is not merely the saddest of diseases, inasmuch as it deprives the patients of the use of their limbs and sooner or later reduces them to almost absolute inaction; it is also a cruel affliction because of the unpleasant sensations which the sufferers experience." Among the unpleasant sensations, he described an "*habitual sensation of excessive heat*" a hitherto undescribed symptom, and the cramps, the feeling of restlessness and uneasiness, of which he gives a vivid and pathetic description (Charcot, 1877, pp. 129–56).

Trousseau had spoken of autopsies performed by Parkinson, Oppolzer, and Lebert on cases of paralysis agitans. To this list Charcot added those performed by Bamberger and Skoda. The findings in the cases reported on by Lebert, Bamberger, and Skoda, he thought, were those of disseminated sclerosis, while those of Parkinson and Oppolzer were probably due to paralysis agitans. "In Parkinson's case," he said, "which by the way HE RELATES AT SECOND HAND, it appears that there was an augmentation of volume, with induration of the pons varolii, of the medulla oblongata, and of the cervical portion of the cord, and that in addition to this the nerves of the tongue and those of the arm were apparently tendinous." He dismissed Oppolzer's findings as inconclusive and ended by asserting that "The foregoing considerations show that the special lesion of paralysis agitans remains to be discovered" (Charcot, 1877, pp. 129–56).

Grasset and Rauzier, in their book, *Traité Pratique des Maladies du Système Nerveux* (Grasset and Rauzier, 1894), confirm that Charcot gave the appellation PARKINSON'S DISEASE to paralysis agitans. "La description particulière de la paralysie agitante de cette espèce à part de tremblement, de cette nevrose spéciale, est due à Parkinson (1817), d'où le nom *maladie de Parkinson*, qui Charcot a proposé de lui donner." An explanation of Charcot's choice of Parkinson's disease in preference to paralysis agitans is given, in these words: "Il y a des cas frustes, dans lesquels le tremblement est reduit, limité aux doigts par exemple, il peut meme manqué entièrement. Le diagnostic se fait alors par l'attitude de sujet, la rigidité musculaire, les sensations de chaleur etc. C'est à cause de l'existence de ces cas frustes que Charcot préfère le nom de *maladie de Parkinson*, qui ne préjuge rien, à celui de paralysie agitante, qui semble impliquer l'agitation, le tremblement" (Grasset and Rauzier, 1894).

British writers were slow in taking the eponym "Parkinson's disease" as an alternative for paralysis agitans. The first time it was mentioned in a textbook published in Great Britain was in 1877 in a book written by Dr. Julius Althaus, *Diseases of the Nervous System* (Althaus, 1877).

It was next mentioned by Dr. Pye-Smith in an edition of Fagge's *Principles and Practice of Medicine* (Fagge 1886), in which a footnote to a description of paralysis agitans states, "abroad it is commonly known as Parkinson's disease." Dr. (later Sir) William Gowers, in 1888, mentioned Parkinson's disease in his classic *Manual of the Diseases of the Nervous System* (Gowers, 1886–88), when he wrote, "From the fact that it was first fully described by Parkinson in 1817, it has been called Parkinson's disease, but the name which he gave to it of 'shaking palsy' is both apt and adequate." In a footnote he added: "To the name 'paralysis agitans' it has been objected by Charcot that either the weakness or tremor is sometimes slight and occurs late, but this fact does not lessen the applicability of the name, since in the majority of cases both symptoms are conspicuous" (Gowers, 1886–88, Vol II. p. 589). Two textbooks on medicine, one by Dr. Frederick Roberts (1839–1918) and the other by Dr. John S. Bristowe (1827–1895), whose later editions both published in 1890 and were popular with the medical students of the day, did not even mention Parkinson or his disease. By the beginning of the 20th century most textbooks of medicine or nervous diseases in particular, invariably mentioned Parkinson's disease.

Articles in the medical press then referred to Parkinson's disease, an early example being a case of post hemiplegic paralysis agitans, by Dr. George F. Johnson, in which he stated, "All the classical symptoms of Parkinson's disease are present."

We have already noted how Thomas Buzzard praised his countryman. He went on to confirm Charcot's original observations on newly observed symptoms in Parkinson's disease, namely rigidity and habitual sensations of excessive heat. Charcot, he said, had also noticed another characteristic; the fixed look and expressionless face. "Far from trembling," said Charcot, "the muscles of the face are motionless, there is a remarkable fixity of look, and the features present a permanent expression of mournfulness, sometimes of stolidness and stupidity."

A symptom, not previously observed, was noted by Buzzard, when he wrote,

> There is a symptom in the patient which according to my experience is a very frequent one, but which has not hitherto, so far as I know, received attention at the hands of any observer. Charcot, it is true, has described the affection of speech, which is often observed, in these words, "There is no real difficulty of speech, but the utterance is slow, jerky, and short of phrase; the pronunciation of each word appears to cost a considerable effort of the will . . ."
>
> We observe something of the kind in this patient, but the symptom upon which I now wish to speak concerns not the articulation but the voice. The voice has a peculiar *piping* character. It is the conventional voice of the old man upon the stage. We all know that on the stage peculiarities of every kind are obliged to be accentuated for the sake of strong light and shade. When a very advanced period of old age has to be represented, the tone of voice adopted is . . . of the shrill *piping* character, which we note in this woman, who however in only 62 years of age.
>
> Shakespeare thus refers to this character of voice in connection with the sixth of the Seven Ages of Man:

'His big manly voice
Turning again towards childish treble pipes
And whistles in his sound'
(Buzzard, 1882, pp. 333–34).

After about 1870, reports of cases of Parkinson's disease and the results of autopsies started appearing in the London medical journals, an outcome probably of the interest taken in the disease by such medical men of international repute as Trousseau and Charcot, whose works had been translated into English and published in London.

Dr. C. Murchison reported the findings of an autopsy to the Pathological Society of London, on May 2, 1871. The case was a man aged 71 who died of typhus, and who had suffered from paralysis agitans for 12 years. He reported that Dr. W. Cayley had made an examination of the spinal cord, and had detected a great increase of the connective tissue outside the cord, which passed upwards at certain points. The central canal was obliterated by a thick mass of rounded cells and granules, and there appeared to be here and there patches of exudate matter (Murchison and Cayley, 1871, pp. 24–27).

Dr. John Murray had a case of advanced paralysis agitans under his care at the Middlesex Hospital, which he reported in the *Lancet* in 1871. A man, aged 62, a bootmaker, had the disease for seven years. The symptoms were very marked, the muscles of his extremities being constantly in motion. Speech was much affected, utterance being almost unintelligible. His progress was characteristic of an advanced stage of the disease, presenting as it did the appearance of falling forwards, which he would actually have done had he not occasionally broken into a shuffling trot. While standing his gait was awkward, his body and arms hanging forwards. Dr. Murray concurred in the belief that the changes are probably always due to chronic connective tissue changes in the cord as described by Dr. Cayley. The patient had been under treatment for several months, and had been given cod liver oil and carbonate of iron, during which time the symptoms were, if anything, less marked.

Dr. Dyce Duckworth wrote a very full account of a case of unilateral paralysis agitans which was under his care at St. Bartholomew's Hospital in 1878. The patient W.K., aged 60, had a constant tremor of the right forearm and hand; paroxysms of trembling awakened him at night; sensation of heat in limb; his writing was quite illegible; the tremors ceased when asleep; and the flexor muscles of the forearm were somewhat contracted and rigid. Dr. Duckworth treated him with bromide of camphor, on Charcot's recommendation, which was followed by some relief and better nights. (Charcot had a few years previously tried hyosyamine and though he said it was not in any sense a cure, it afforded considerable relief. He mentioned Ergot, Belladonna, nitrate of silver, arsenate of potash, bromide of camphor, and carbonate of soda, all of which had been tried but found wanting.) Of the pathology of the disease Dr. Duckworth noted that, "morbid anatomy throws no light upon the nature of the disease, or, at all events shows no definite and constantly recurring lesion" (Duckworth, 1879, p. 342). Later the patient developed a commencing tremor in the left arm and hand.

A case at St. Thomas's Hospital, under the care of Dr. Ord, was reported in the *British Medical Journal* in 1881. Dr. Hadden, who reported the case for Dr. Ord, stated that he had examined the reflexes in this case and had noted, "marked exaggeration of tendon reflexes in the upper and lower extremities certainly suggest sclerosis affecting probably the lateral columns of the cord" (Hadden, 1881, p. 936).

Even at this late date, the neurologist had to rely on the opinions regarding the pathologists Parkinson, Oppolzer, Bamberger, Lebert, and Skoda. With reference to the pathology of the disease, Dr. Hadden said, "Charcot's investigations negative the idea that paralysis agitans is accompanied by hyperplasis of the connective tissue, such as is seen in disseminated sclerosis. In the cases of Bamberger, Lebert and Skoda, distinct evidence of such a degeneration are forthcoming, but the changes have been considered by Charcot as distinctive of disseminated sclerosis. Marshall Hall, Parkinson and Oppolzer, have described changes in the pons, medulla and cord in their cases. It is possible that in some cases the lateral columns become affected secondarily, and, hence, the occurrence of the rigidity and cramps so frequently seen in these cases" (Hadden, 1881, pp. 936–37).

Dr. C.E. Beevor in 1885 described four cases of paralysis agitans, diagnosed in the absence of tremors. Charcot had described two, Gowers and Buzzard one each.

The *Lancet* in 1885 had received plaintive appeals for suggestions on how to treat the disease, from desperate general practitioners, and the following letter was written in reply. ". . . relief and marked benefit have resulted from the administration of the extract of physostigna in doses increased up to ½ gr. five times a day, with 5 grs. of compound as safoetida in pill form. The insomnia has been relieved, the tongue cleansed, the tremor lessened, and the despondency lightened. Conium juice in doses up to two ounces six times a day was of no avail. A.H. Bampton, M.D." (Bampton, 1885, p. 189).

The treatment of paralysis agitans received attention in the *British Medical Journal* in 1887. Subcutaneous injections were now in vogue, and morphine and curare given in this manner were recommended. Arsenic given subcutaneously, in the form of m.vi to x of Fowlers solution in distilled water, was also in favour. By mouth, calabar bean, chloral hydrate, potassium bromide, and silver nitrate, it was suggested, were worth trial. Dr. Russell Reynolds still recommended electricity via the Pulermacher chain (an instrument producing electrical discharges), despite Charcot's declaration of its inutility (Alexander, 1887, p. 1085).

The pathology of Parkinson's disease was the subject of an editorial in the *Lancet* in 1888. M. Teisser (Lyon Médical No. 28) had submitted many arguments in favour of the view that paralysis agitans is really an organic disease of the spinal cord. "A diffuse sclerosis of the lateral columns had been found, in some cases extending up the vesicular column of Clarke and into the intermedio-lateral tract." It was considered that "The main conclusion to be drawn, if M. Teisser's observations are exact, seems to be that paralysis agitans is, like chorea, a symptom, and not a disease in itself" (*Lancet*, 1888, p. 130).

An abstract of an article by von Ketscher appeared in the *Lancet* in 1893, giving an account of post mortem examinations on three cases of paralysis agitans.

The cases of paralysis agitans with reference to their pathological anatomy are divided into two classes—those in which the examination has furnished negative results, and those in which various changes have been found in the nervous system, such as hyperplasia of the connective tissue and neuroglia of the spinal cord, alteration in the nervous system itself and vascular changes. Similar changes have been described as occurring in the medulla and pons. Three cases of undoubted Parkinson's disease have been investigated by von Ketscher. The nervous system, central and peripheral, were examined, and in all three cases changes were found in these regions. The nervous structures showed atrophy, the ganglion cells of the brain were deeply pigmented, and altered in form, the nerve fibres, both in the peripheral nerves and in the spinal cord were degenerated and had in some instances disappeared, whilst the muscular fibres also were atrophied and degenerated. The neuroglia was thickened, especially around the vessels, and mostly in the posterior and lateral columns. The vessels also were altered, their walls thickened, and there were miliary aneurysms and small haemorrhages present. Similar changes but slighter in degree, were found as senile changes in patients not the subjects of paralysis agitans; von Ketscher concludes, with Borgherini and others, that paralysis agitans is only the expression of an extreme and premature senility of the nervous system, and he is of opinion that the primary changes are in the vessels, those in the nervous system being secondary (*Lancet*, 1893, pp. 669–70).

The position at the end of the 19th century was unchanged, it was still thought that the proximate cause, as Parkinson termed it, was in the spinal cord and medulla oblongata. One more drug was added to the long list of remedies when Dr. Sacaze, of Montpelier, claimed a remarkable success by the administration of sodium borate, but this was an isolated case (*Lancet*, 1893, p. 377).

The incidence of Parkinson's disease was occasionally referred to by various observers. Charcot believed that the disease was commoner in Britain and North America, when he said, "There is some reason to believe that the Anglo-Saxon race is most subject to the disease." At the same time he remarked, "even today I do not believe disseminated sclerosis is known in England." The proportion in which the different sexes were affected was more often referred to. Charcot said it occurred as often in females as males, while Dr. W.R. Sanders said that between the years 1855–1863 some 205 persons died of the disease in England and Wales, of whom on average, per annum 14 were males and 8 females. Pye-Smith said that between the years 1866–1875, some 14 cases had been admitted into the wards of Guy's Hospital, of which 12 were males and only 2 females. It will be remembered that Parkinson's six cases were all males. (Sir) William Gowers (1888) had seen 80 cases of whom 50 were males, 30 females. Dr. Frederick Petersen of New York (1891) had made an analysis of 47 cases, 29 males and 18 females. Dr. Julius Althaus said that in the year 1874 there were 69 fatal cases of the disease in England and Wales, 42 males and 27 females. The consensus of opinion, therefore, was that it was commoner in males. Pye-Smith said that 6 out of the 14 cases commenced between the ages of 50 and 59; Gowers said 50% of his cases began between the ages of 50 and 60; and Parkinson's cases also started between 50 and 60. Dr. Petersen also found the most common age of onset was between 50 and 60.

We can thus see that the syndrome so brilliantly described by Parkinson had attracted some of the best minds of 19th century medicine and neurology. Several features of its course and incidence had been added, and its neuropathology and treatment had exercised many. But a further understanding of the genesis and control of this "saddest of diseases" had to wait until the present century.

Appendix

An Essay on the Shaking Palsy

(Original Text)

JAMES PARKINSON
MEMBER OF THE ROYAL COLLEGE OF SURGEONS

Preface

The advantages which have been derived from the caution with which hypothetical statements are admitted, are in no instance more obvious than in those sciences which more particularly belong to the healing art. It therefore is necessary, that some conciliatory explanation should be offered for the present publication: in which, it is acknowledged, that mere conjecture takes the place of experiment; and, that analogy is the substitute for anatomical examination, the only sure foundation for pathological knowledge.

When, however, the nature of the subject, and the circumstances under which it has been here taken up, are considered, it is hoped that the offering of the following pages to the attention of the medical public, will not be severely censured. The disease, respecting which the present inquiry is made, is of a nature highly afflictive. Notwithstanding which, it has not yet obtained a place in the classification of nosologists; some have regarded its characteristic symptoms as distinct and different diseases, and others have given its name to diseases differing essentially from it; whilst the unhappy sufferer has considered it as an evil, from the domination of which he had no prospect of escape.

The disease is of long duration: to connect, therefore, the symptoms which occur in its later stages with those which mark its commencement, requires a continuance of observation of the same case, or at least a correct history of its symptoms, even for several years. Of both these advantages the writer has had the opportunities of availing himself; and has hence been led particularly to observe several other cases in which the disease existed in different stages of its progress. By these repeated observations, he hoped that he had been led to a probable conjecture as to the nature of the malady, and that analogy had suggested such means as might be productive of relief, and perhaps even of cure, if employed before the disease had been too long established. He therefore considered it to be a duty to submit his opinions to the examination of others, even in their present state of immaturity and imperfection.

To delay their publication did not, indeed, appear to be warrantable. The disease had escaped particular notice; and the task of ascertaining its nature and cause by anatomical investigation, did not seem likely to be taken up by those who, from their abilities and opportunities, were most likely to accomplish it. That these friends to humanity and medical science, who have already unveiled to us many of the morbid processes by which health and life is abridged, might be excited to extend their researches to this malady, was much desired; and it was hoped, that this might be procured by the publication of these remarks.

Should the necessary information be thus obtained, the writer will repine at no censure which the precipitate publication of mere conjectural suggestions may incur; but shall think himself fully rewarded by having excited the attention of those, who may point out the most appropriate means of relieving a tedious and most distressing malady.

Chapter I. Definition—History—Illustrative Cases

Shaking Palsy (Paralysis Agitans)

Involuntary tremulous motion, with lessened muscular power, in parts not in action and even when supported; with a propensity to bend the trunk forward, and to pass from a walking to a running pace: the senses and intellects being uninjured.

The term Shaking Palsy has been vaguely employed by medical writers in general. By some it has been used to designate ordinary cases of Palsy, in which some slight tremblings have occurred; whilst by others it has been applied to certain anomalous affections, not belonging to Palsy.

The shaking of the limbs belonging to this disease was particularly noticed, as will be seen when treating of the symptoms, by Galen, who marked its peculiar character by an appropriate term. The same symptom, it will also be seen, was accurately treated of by Sylvius de la Boe. Juncker also seems to have referred to this symptom: having divided tremor into active and passive, he says of the latter, "ad affectus semiparalyticos pertinant; de qualibus hic agimus, quique *tremores paralytoidei* vocantur." Tremor has been adopted, as a genus, by almost every nosologist; but always unmarked, in their several definitions, by such characters as would embrace this disease. The celebrated Cullen, with his accustomed accuracy observes, "Tremorem, utpote semper symptomaticum, in numerum generum recipere nollem; species autem a Sauvagesio recensitas, prout mihi vel astheniaevel paralysios, vel convulsionis symptomata esse videntur, his subjungam."* Tremor can indeed only be considered as a

*Synopsis Nosologiae Methodicae.—Tom. ii. p. 195.

symptom, although several species of it must be admitted. In the present instance, the agitation produced by the peculiar species of tremor, which here occurs, is chosen to furnish the epithet by which this species of Palsy, may be distinguished.

History

So slight and nearly imperceptible are the first inroads of this malady, and so extremely slow is its progress, that it rarely happens, that the patient can form any recollection of the precise period of its commencement. The first symptoms perceived are, a slight sense of weakness, with a proneness to trembling in some particular part; sometimes in the head, but most commonly in one of the hands and arms. These symptoms gradually increase in the part first affected; and at an uncertain period, but seldom in less than twelvemonths or more, the morbid influence is felt in some other part. Thus assuming one of the hands and arms to be first attacked, the other, at this period becomes similarly affected. After a few more months the patient is found to be less strict than usual in preserving an upright posture: this being most observable whilst walking, but sometimes whilst sitting or standing. Sometime after the appearance of this symptom, and during its slow increase, one of the legs is discovered slightly to tremble, and is also found to suffer fatigue sooner than the leg of the other side: and in a few months this limb becomes agitated by similar tremblings, and suffers a similar loss of power.

Hitherto the patient will have experienced but little inconvenience; and befriended by the strong influence of habitual endurance, would perhaps seldom think of his being the subject of disease, except when reminded of it by the unsteadiness of his hand, whilst writing or employing himself in any nicer kind of manipulation. But as the disease proceeds, similar employments are accomplished with considerable difficulty, the hand failing to answer with exactness to the dictates of the will. Walking becomes a task which cannot be performed without considerable attention. The legs are not raised to that height, or with that promptitude which the will directs, so that the utmost care is necessary to prevent frequent falls.

At this period the patient experiences much inconvenience, which unhappily is found daily to increase. The submission of the limbs to the directions of the will can hardly ever be obtained in the performance of the most ordinary offices of life. The fingers cannot be disposed of in the proposed directions, and applied with certainty to any proposed point. As time and the disease proceed, difficulties increase: writing can now be hardly at all accomplished; and reading, from the tremulous motion, is accomplished with some difficulty. Whilst at meals the fork not being duly directed frequently fails to raise the morsel from the plate: which, when seized, is with much difficulty conveyed to the mouth. At this period the patient seldom experiences a suspension of the agitation of his limbs.

Commencing, for instance in one arm, the wearisome agitation is borne until beyond sufferance, when by suddenly changing the posture it is for a time stopped in that limb, to commence, generally, in less than a minute in one of the legs, or in the arm of the other side. Harassed by this tormenting round, the patient has recourse to walking, a mode of exercise to which the sufferers from this malady are in general partial; owing to their attention being thereby somewhat diverted from their unpleasant feelings, by the care and exertion required to ensure its safe performance.

But as the malady proceeds, even this temporary mitigation of suffering from the agitation of the limbs is denied. The propensity to lean forward becomes invincible, and the patient is thereby forced to step on the toes and fore part of the feet, whilst the upper part of the body is thrown so far forward as to render it difficult to avoid falling on the face. In some cases, when this state of the malady is attained, the patient can no longer exercise himself by walking in his usual manner, but is thrown on the toes and forepart of the feet; being, at the same time, irresistibly impelled to take much quicker and shorter steps, and thereby to adopt unwillingly a running pace. In some cases it is found necessary entirely to substitute running for walking; since otherwise the patient, on proceeding only a very few paces, would inevitably fall.

In this stage, the sleep becomes much disturbed. The tremulous motion of the limbs occur during sleep, and augment until they awaken the patient, and frequently with much agitation and alarm. The power of conveying the food to the mouth is at length so much impeded that he is obliged to consent to be fed by others. The bowels, which had been all along torpid, now, in most cases, demand stimulating medicines of very considerable power: the expulsion of the faeces from the rectum sometimes requiring mechanical aid. As the disease proceeds towards its last stage, the trunk is almost permanently bowed, the muscular power is more decidedly diminished, and the tremulous agitation becomes violent. The patient walks now with great difficulty, and unable any longer to support himself with his stick, he dares not venture on this exercise, unless assisted by an attendant, who walking backwards before him, prevents his falling forwards, by the pressure of his hands against the fore part of his shoulders. His words are now scarcely intelligible; and he is not only no longer able to feed himself, but when the food is conveyed to the mouth, so much are the actions of the muscles of the tongue, pharynx, &c. impeded by impaired action and perpetual agitation, that the food is with difficulty retained in the mouth until masticated; and then as difficulty swallowed. Now also, from the same cause, another very unpleasant circumstance occurs: the saliva fails of being directed to the back part of the fauces, and hence is continually draining from the mouth, mixed with the particles of food, which he is no longer able to clear from the inside of the mouth.

As the debility increases and the influence of the will over the muscles fades away, the tremulous agitation becomes more vehement. It now seldom leaves him for a moment; but even when exhausted nature seizes a small portion of sleep, the motion becomes so violent as not only to shake the bed-hangings, but even the floor and sashes of the room. The chin is now almost immoveably bent down upon the sternum. The slops with which he is attempted to be fed, with the saliva, are continually trickling from the mouth. The power of articulation is lost. The urine and faeces are passed involuntarily; and at the last, constant sleepiness, with slight delirium, and other marks of extreme exhaustion, announce the wished-for release.

Case I

Almost every circumstance noted in the preceding description, was observed in a case which occurred several years back, and which, from the particular symptoms which manifested themselves in its progress; from the little knowledge of its nature, acknowledged to be possessed by the physician who attended; and from the mode of its termination; excited an eager wish to acquire some further knowledge of its nature and cause.

The subject of this case was a man rather more than fifty years of age, who had industriously followed the business of a gardener, leading a life of remarkable temperance and sobriety. The commencement of the malady was first manifested by a slight trembling of the left hand and arm, a circumstance which he was disposed to attribute to his having been engaged for several days in a kind of employment requiring considerable exertion of that limb. Although repeatedly questioned, he could recollect no other circumstance which he could consider as having been likely to have occasioned his malady. He had not suffered much from Rheumatism, or been subject to pains of the head, or had ever experienced any sudden seizure which could be refered to apoplexy or hemiplegia. In this case, every circumstance occurred which has been mentioned in the preceding history.

Case II

The subject of the case which was next noticed was casually met with in the street. It was a man sixty-two years of age; the greater part of whose life had been spent as an attendant at a magistrate's office. He had suffered from the disease about eight or ten years. All the extremities were considerably agitated, the speech was very much interrupted, and the body much bowed and shaken. He walked almost entirely on the fore part of his feet, and would have fallen every step if he had not been supported by his stick. He described the disease as having come on very gradually, and as being, according to his full assurance, the consequence of considerable irregularities in his mode of living, and particularly of indulgence in spirituous liquors. He was the inmate of a poor-house of a distant parish, and being fully assured of the incurable nature of his complaint, declined making any attempts for relief.

Case III

The next case was also noticed causally in the street. The subject of it was a man of about sixty-five years of age, of a remarkable athletic frame. The agitation of the limbs, and indeed of the head and of the whole body, was too vehement to allow it to be designated as trembling. He was entirely unable to walk; the body being so bowed, and the head thrown so forward, as to oblige him to go on a continued run, and to employ his stick every five or six steps to force him more into an upright posture, by projecting the point of it with great force against the pavement. He stated, that he had been a sailor, and attributed his complaints to having been for several months confined in a Spanish prison, where he had, during the whole period of his confinement, lain upon the bare damp earth. The disease had here continued so long, and made such a progress, as to afford little or no prospect of relief. He besides was a poor mendicant, requiring as well as the means of medical experiment, those collateral aids which he could only obtain in an hospital. He was therefore recommended to make trial if any relief could, in that mode, be yielded him. The poor man, however, appeared to be by no means disposed to make the experiment.

Case IV

The next case which presented itself was that of a gentleman about fifty-five years, who had first experienced the trembling of the arms about five years before. His application was on account of a considerable degree of inflammation over the lower ribs on the left side, which terminated in the formation of matter beneath the fascia. About a pint was removed on making the necessary opening; and a considerable quantity discharged daily for two or three weeks. On his recovery from this, no change appeared to have taken place in his original complaint; and the opportunity of learning its future progress was lost by his removal to a distant part of the country.

Case V

In another case, the particulars of which could not be obtained, and the gentleman, the lamented subject of which was only seen at a distance, one of the characteristic symptoms of this malady, the inability for motion, except in a running pace, appeared to exist in an extraordinary degree. It seemed to be necessary that the gentleman should be supported by his attendant, standing before him with a hand placed on each shoulder, until, by gently swaying backward and forward, he had placed himself in equipoise; when, giving the word, he would start in a running pace, the attendant sliding from before him and running forward, being ready to receive him and prevent his falling, after his having run about twenty paces.

Case VI

In a case which presented itself to observation since those above-mentioned, every information as to the progress of the malady was very readily obtained. The gentleman who was the subject of it is seventy-two years of age. He has led a life of temperance, and has never been exposed to any particular situation or circumstance which he can conceive likely to have occasioned, or disposed to this complaint; which he rather seems to regard as incidental upon his advanced age, than as an object of medical attention. He however recollects, that about twenty years ago, he was troubled with lumbago, which was severe and lasted some time. About eleven or twelve, or perhaps more, years ago, he first perceived weakness in the left hand and arm, and soon after found the trembling commence. In about three years afterwards the right arm became affected in a similar manner: and soon afterwards the convulsive motions affected the whole body, and began to interrupt the speech. In about three years from that time the legs became affected. Of late years the action of the bowels had been very much retarded; and at two or three different periods had, with great difficulty, been made to yield to the action of very strong cathartics. But within the last twelvemonths this difficulty has not been so great; perhaps owing to an increased secretion of mucus, which envelopes the passing faeces, and which precedes and follows their discharge in considerable quantity.

About a year since, on waking in the night, he found that he had nearly lost the use of the right side, and that the face was much drawn to the left side. His medical attendant saw him the following day, when he found him languid, with a small and quick pulse, and without pain in the head or disposition to sleep. Nothing more therefore was done than to promote the action of the bowels, and apply a blister to the back of the neck, and in about a fortnight the limbs had entirely recovered from their palsied state. During the time of their having remained in this state, neither the arm nor the leg of the paralytic side was in the least affected with the tremulous agitation; but as their paralysed state was removed, the shaking returned.

At present he is almost constantly troubled with the agitation, which he describes as generally commencing in a slight degree, and gradually increasing, until it arises to such a height as to shake the room; when, by a sudden and somewhat violent change of posture, he is almost always able to stop it. But very soon afterwards it will commence in some other limb, in a small degree, and gradually increase in violence; but he does not remember the thus checking of it, to have been followed by any injurious effect. When the agitation had not been thus interrupted, he stated, that it gradually extended through all the limbs, and at last affected the whole trunk. To illustrate his observation as to the power of suspending the motion by a sudden change of posture, he, being then just come in from a walk, with every limb shaking, threw himself rather violently into a chair, and said, "Now I am as well as ever I was in my life." The shaking completely stopped; but returned within two minutes' time.

He now possessed but little power in giving a required direction to the motions of any part. He was scarcely able to feed himself. He had written hardly intelligibly for the last three years; and at present could not write at all. His attendants observed, that of late the trembling would sometimes begin in his sleep, and increase until it awakened him: when he always was in a state of agitation and alarm.

On being asked if he walked under much apprehension of falling forwards? he said he suffered much from it; and replied in the affirmative to the question, whether he experienced any difficulty in restraining himself from getting into a running pace? It being asked, if whilst walking he felt much apprehension from the difficulty of raising his feet, if he saw a rising pebble in his path? he avowed, in a strong manner, his alarm on such occasions; and it was observed by his wife, that she believed, that in walking across the room, he would consider as difficulty the having to step over a pin.

The preceding cases appear to belong to the same species: differing from each other, perhaps, only in the length of time which the disease had existed, and the stage at which it had arrived.

Chapter II. Pathognomic Symptoms Examined—*Tremor Coactus—Scelotyrbe Festinians*

It has been seen in the preceding history of the disease, and in the accompanying cases, that certain affections, the tremulous agitations, and the almost invincible propensity to run, when wishing only to walk, each of which has been considered by nosologists as distinct diseases, appear to be pathognomic symptoms of this malady. To determine in which of these points of view these affections ought to be regarded, an examination into their nature, and an inquiry into the opinions of preceding writers respecting them, seem necessary to be attempted.

I. *Involuntary tremulous motion, with lessened voluntary muscular power, in parts, not in action, and even supported.*

It is necessary that the peculiar nature of this tremulous motion should be ascertained, as well for the sake of giving to it its proper designation, as for assisting in forming probable conjectures, as to the nature of the malady, which it helps to characterise. Tremors were distinguished by Juncker into Active, those proceeding from sudden affection of the minds, as terror, anger, &c. and Passive, dependant on debilitating causes, such as advanced age, palsy, &c.* But a much more satisfactory and useful distinction is made by Sylvius de la Boë into those tremors which are produced by attempts at voluntary motion, and those which occur whilst the body is at rest.** Sauvages distinguishes the latter of these species (*Tremor Coactus*) by observing, that the tremulous parts leap, and as it were vibrate, even when supported: whilst every other tremor, he observes, ceases, when the voluntary exertion for moving the limbs stops, or the part is supported, but returns when we will the limb to move; whence, he says, tremor is distinguished from every other kind of spasm.†

A small degree of attention will be sufficient to perceive, that Sauvages, by this just distinction, actually separates this kind of tremulous motion, and which is the kind peculiar to this disease, from the Genus Tremor. In doing this he is fully warranted by the observations of Galen on the same subject, as noticed by Van Swieten.‡ "Binas has tremoris species§ Galenus subtiliter distinxit, atque etiam diversis nominibus insignivit, tremor enim (τρβμδ) facultatis corpus moventis et vehentis infirmitate oboritur. Quippe nemo, qui artus movere non instituerit tremet. Palpitantes autem partes, etiam in quiete fuerint, etiamsi nullum illis motum induxeris palpitant. Ideo primam (*posteriorem*) modo descriptam tremoris speciem, quando quiescenti homini involuntariis illis et alternis motibus agitantur membra, palpitationem (παλμον) dixit, posteriorem (*primam*) vero, quae non fit nisi homo conetur partes quasdam movere tremorem vocavit."

---

*Junckeri conspect. de tremore.

**Sect. v. Ubi autem solito pauciores deferunter ad eadem organa spiritus animales, imperfectae ac imbecillae observantur fieri eadem functiones, in motu tremulo et infirmo, nec diu durante, in visu debili, ac mox defatigato, &c.

Sect. XIX. Inaequaliter, inordinatè, ac praeter contraque voluntatem moventur spiritus animales per nervos ad partes mobiles, in motu convulsivo, ac tremore, quassuve membrorum coacto.

Distinguendus namque his tremor quiescente licet ac decumbente corpore molustus a motu tremulo, de quo dictum. Sect. v. Quique quiescente corpore cessat, eodemque iterum moto repetit.

Sect. XXV. Coactus tremor debetur animalibus spiritibus inordinatè ac continuo, cum aliquo impetu ad trementium membrorum musculos per nervos propulsis: sive fuerit is universalis, sive particularis, sive corpus fuerit ad huc robustum sive debile, Sylvii de la Boë. Prax. lib i. cap. xlii.

†Nosolog. Methodic. Auctore Fr. Boissier de Sauvages, Tomi. II. Partis ii. p. 54. 1763.

‡Comment. in Herman. Boerhaav. Aphorismos. Tom. ii. p. 181.

§De tremore. Cap. 3 and 4. Chart, Tom. vii. p. 200 201.

Under this authority the term palpitation may be employed to mark those morbid motions which chiefly characterise this disease, notwithstanding that this term has been anticipated by Sauvages, as characteristic of another species of tremor.* The separation of palpitation of the limbs (*Palmos* of Galen, *Tremor Coactus* of de la Boë) from tremor, is the more necessary to be insisted on, since the distinction may assist in leading to a knowledge of the seat of the disease. It is also necessary to bear in mind, that this affection is distinguishable from tremor, by the agitation, in the former, occurring whilst the affected part is supported and unemployed, and being even checked by the adoption of voluntary motion; whilst in the latter, the tremor is induced immediately on bringing the parts into action. Thus an artist, afflicted with the malady here treated of, whilst his hand and arm is palpitating strongly, will seize his pencil, and the motions will be suspended, allowing him to use it for a short period; but in tremor, if the hand be quite free from the affection, should the pen or pencil be taken up, the trembling immediately commences.

II. *A propensity to bend the trunk forwards, and to pass from a walking to a running pace.*

This affection, which observation seems to authorise the being considered as a symptom peculiar to this disease, has been mentioned by few nosologists: it appears to have been first noticed by Gaubius, who says, "Cases occur in which the muscles duly excited into action by the impulse of the will, do then, with an unbidden agility, and with an impetus not to be repressed, accelerate their motion, and run before the unwilling mind. It is a frequent fault of the muscles belonging to speech, nor yet of these alone: I have seen one, who was able to run, but not to walk.†"

---

*Sect. XVI. *Tremor palpitans*, Preysinger classis morborum. *Palmos* Galeni.

In tremoribus vulgaribus, aequalibus temporum intervallis, non musculus, sed artus ipsemet alternatim attollitur aut deprimitur, aut in oppositas partes it atque redit per minima tamen spatiola; in palpitatione verè sine ullo ordine musculi unius lacertus subito subsilit, nec regulariter continuoque movetur, sed nunc semel aut bis, nunc minimè intra idem tempus subsilit; an causa irritans in sensorio communi, an in musculo ipse palpitante Quaerenda sit, ignoramus. *Nosologiae Methodicae*, Vol. I. p. 559. 1768.

But the adoption which Sauvages has made of this term, will not be regarded as an absolute prohibition from the employment of its here; since the *tremor palpitans* of Sauvages should be considered rather as a palpitation of the muscles, whilst the motion which is so prominent a symptom in this disease, may be considered as a palpitation of the limbs.

†Est et ubi musculi, recte quidem ad voluntatis nutum in actum concitati, injussa dein agilitate atque impetu non reprimendo motus suos accelerant, mentemque invitam praecurrunt. Vitium loquelae musculis frequens, nec his solis tamen proprium: vidi enim, qui currere, non gradi, poterat.*

*Institution. Patholog. Medicinal. Auctore. H.D. Gaubio. 751.

Sauvages, referring to this symptom, says, another disease which has been very rarely seen by authors, appears to be referable to the same genus (Scelotyrbe, of which he makes *Chorea sancti viti* the first species); which, he says, "I think cannot be more fitly named than hastening or hurrying Scelotyrbe (*Scelotyrbem festinantem, seu festiniam*)."

Scelotyrbe festinans, he says, is a peculiar species of scelotyrbe, in which the patients, whilst wishing to walk in the ordinary mode, are forced to run, which has been seen by Carguet and by the illustrious Gaubius; a similar affection of the speech, when the tongue thus outruns the mind, is termed volubility. Mons. de Sauvages attributes this complaint to a want of flexibility in the muscular fibres. Hence, he supposes, that the patients make shorter steps, and strive with a more than common exertion or impetus to overcome the resistance; walking with a quick and hastened step, as if hurried along against their will. *Chorea Viti*, he says, attacks the youth of both sexes, but this disease only those advanced in years; and adds, that it has hitherto happened to him to have seen only two of these cases; and that he has nothing to offer respecting them, either in theory or practice.*

Having made the necessary inquiries respecting these two affections, *Tremor coactum* of Sylvius de la Boë and of Sauvages, and *Scelotyrbe festinans* of the latter nosologist, which appear to be characteristic symptoms of this disease, it becomes necessary, in the next place, to endeavour to distinguish this disease from others which may bear a resemblance to it in some particular respects.

---

*Ad idem genus morbi altera species rarissima ab auctoribus praetervisa referenda videtur, quam non aptius nominari posse putem quàm scelotyrbem festinantem, seu festiniam.

Sect. II. *Scelotyrbe festinans*; est peculiaris scelotyrbes species in qua aegri solito more dum gradi volunt currere cognuntur, quod videre est aud D. Carquet, et observavit Leydae illustr. Gaubius. *Patholog. instit.* 751, et in loquela haec *volubilitas* dicitur quâ lingua praecurit mentem. Video actu mulierem sexagenariam hoc affectam morbo siccitati nervorum tribuendo; laborat enim rheumatismo sicco, seu ab acrimonia sanguinis, dolores nocte a calore recrudescunt, à thermis non sublevantur: ei praescripsi phlebotomiam, et praemissis jusculis ex lactucâ, endiviâ, et collo arietis, lene catharticum, inde vero lacticinia.

Est affinitas cum scelotyrbe, chorea viti, deest flexibilitas in fibris musculorum; unde motus breves edunt, et conatu seu impetu solito majori, cum resistentiam illam superare nitunter, velut inviti festinant, ac praecipiti seu concitato passu gradiuntur. Chorea viti pueros, puellasve impuberes aggreditur; festinia vero senes, et duos tantum hactenus observare mihi contigit. Quam multos autem videmus morbos, paucissimosque observamus. De theorià et pràxi nihil habeo quod dicam; etenim sola experienta praxim cujusvis morbi determinat, et ex hac pro felici vel infausto successu theoria dein elicienda est. *Nosolog. Methodic.* Auctore, Fr. Boissier de Sauvages. Tomi. II. Part ii. p. 108.

Chapter III. Shaking Palsy Distinguished from Other Diseases With Which it May Be Confounded

Treating of a disease resulting from an assemblage of symptoms, some of which do not appear to have yet engaged the general notice of the profession, particular care is required whilst endeavouring to mark its diagnostic characters. It is sufficient, in general, to point out the characteristic differences which are observable between diseases in some respects resembling each other. But in this case more is required: it is necessary to show that it is a disease which does not accord with any which are marked in the systematic arrangements of nosologists; and that the name by which it is here distinguished has been hitherto vaguely applied to diseases very different from each other, as well as from that to which it is now appropriated.

Palsy, either consequent to compression of the brain, or dependent on partial exhaustion of the energy of that organ, may, when the palsied limbs become affected with tremulous motions, be confounded with this disease. In those cases the abolition or diminution of voluntary muscular action takes place suddenly, the sense of feeling being sometimes also impaired. But in this disease, the diminution of the influence of the will on the muscles comes on with extreme slowness, is always accompanied, and even preceded, by agitations of the affected parts, and never by a lessened sense of feeling. The dictates of the will are even, in the last stages of the disease, conveyed to the muscles; and the muscles act on this impulse, but their actions are perverted.

Anomalous cases of convulsive affections have been designated by the term Shaking Palsy: a term which appears to be improperly applied to these cases, independent of the want of accordance between them and that disease which has been here denominated Shaking Palsy. Dr. Kirkland, in his commentary on Apoplectic and Paralytic Affections, &c. cites the following case, related by Dr. Charlton, as belonging, he says, to the class of Shaking Palsies. "Mary Ford, of a sanguineous and robust constitution, had an involuntary motion of her right arm, occasioned by a fright, which first brought on convulsion fits, and most excruciating pain in the stomach, which vanished on a sudden, and her right arm was instantaneously flung into an involuntary and perpetual motion, like the swing of a pendulum, raising the hand, at every vibration higher than her head; but if by any means whatever it was stopped; the pain in her stomach came on again, and convulsion fits were the certain consequence, which went off when the vibration of her hand returned."

Another case, which the Doctor designates as 'A Shaking Palsy,' apparently from worms, he describes thus, "A poor boy, about twelve or thirteen years of age, was seized with a Shaking Palsy. His legs became useless, and together with his head and hands, were in continual agitation; after many weeks trial of various remedies, my assistance was desired."

"His bowels being cleared, I ordered him a grain of Opium a day in the gum pill; and in three or four days the shaking had nearly left him." By pursuing this plan, the medicine proving a vermifuge, he could soon walk, and was restored to perfect health.

Whether these cases should be classed under Shaking Palsy or not, is necessary to be here determined; since, if they are properly ranked, the cases which have been described in the preceding pages, differ so much from them as certainly to oppose their being classed together: and the disease, which is the subject of these pages, cannot be considered as the same with Shaking Palsy, as characterised by those cases.

The term Shaking Palsy is evidently inapplicable to the first of these cases, which appears to have belonged more properly to the genus *Convulsio*, of Cullen, or to *Hieranosos* of Linnaeus and Vogel.*

The latter appears to be referable to that class of proteal forms of disease, generated by a disordered state of primae viae, sympathetically affecting the nervous influence in a distant part of the body.

Unless attention is paid to one circumstance, this disease will be confounded with those species of passive tremblings to which the term Shaking Palsies has frequently been applied. These are, *tremor temulentus*, the trembling consequent to indulgence in the drinking of spirituous liquors; that which proceeds from the immoderate employment of tea and coffee; that which appears to be dependent on advanced age; and all those tremblings which proceed from the various circumstances which induce a diminution of power in the nervous system. But by attending to that circumstance alone, which has been already noted as characteristic of mere tremor, the distinction will readily be made. If the trembling limb be supported, and none of its muscles be called into action, the trembling will cease. In the real Shaking Palsy the reverse of this takes place, the agitation continues in full force whilst the limb is at rest and unemployed; and even is sometimes diminished by calling the muscles into employment.

---

*Corporis agitatio continua, indolens, convulsiva, cum sensibilitate.—*Linn.*

Agitatio corporis vel artuum convulsiva continua, chronica, cum integritate sensuum.—*Vogel.*

This genus is resolved by Cullen into that of Convulsio. *Synops. Nosol.* 1803.

Dr. Macbride has given a very interesting and illustrative case of this disease.

"Hieranasós, or Morbus Sacer, so called, as being vulgarly supposed to arise from witchcraft, or some extraordinary celestial influence, is a distinct genus of disease, though a very uncommon one; the author once had an opportunity of seeing a case. The patient was a lad about seventeen, who at that time had laboured under this extraordinary disease for more than twelve years. His body was so distorted, and the legs and arms so twisted round it, by the continued convulsive working, that no words can give an adequate idea of the oddity of his figure; the agitation of the muscles was perpetual; but in general he did not complain of pain nor sickness; and had his senses perfectly, insomuch that he used to assist his mother, who kept a little school, in teaching children to read." *A methodical Introduction to the Theory and Practice of Physic. By David Macbride, M.D. p. 559*

Chapter IV. Proximate Cause—Remote Causes—Illustrative Cases

Before making the attempt to point out the nature and cause of this disease, it is necessary to plead, that it is made under very unfavourable circumstances. Unaided by previous inquiries immediately directed to this disease, and not having had the advantage, in a single case, of that light which anatomical examination yields, opinions and not facts can only be offered. Conjecture founded on analogy, and an attentive consideration of the peculiar symptoms of the disease, have been the only guides that could be obtained for this research, the result of which is, as it ought to be, offered with hesitation.

Supposed Proximate Cause

A diseased state of the *medulla spinalis*, in that part which is contained in the canal, formed by the superior cervical vertebrae, and extending, as the disease proceeds, to the *medulla oblongata*.

By the nature of the symptoms we are taught, that the disease depends on some irregularity in the direction of the nervous influence; by the wide range of parts which are affected, that the injury is rather in the source of this influence than merely in the nerves of the parts; by the situation of the parts whose actions are impaired, and the order in which they become affected, that the proximate cause of the disease is in the superior part of the medulla spinalis; and by the absence of any injury to the senses and to the intellect, that the morbid state does not extend to the encephalon.

Uncertainty existing as to the nature of the proximate cause of this disease, its remote causes must necessarily be referred to with indecision. Assuming however the state just mentioned as the proximate cause, it may be concluded that this may be the result of injuries of the medulla itself, or of the theca helping to form the canal in which it is inclosed.

The great degree of mobility in that portion of the spine which is formed by the superior cervical vertebrae, must render it, and the contained parts, liable to injury from sudden distortions. Hence therefore may proceed inflammation of quicker or of slower progress, disease of the vertebrae, derangement of structure in the medulla, or in its membranes, thickening or even ulceration of the theca, effusion of fluids, &c.

But in no case which has been noticed, has the patient recollected receiving any injury of this kind, or any fixed pain in early life in these parts, which might have led to the opinion that the foundation for this malady had been thus laid. On the subject indeed of remote causes, no satisfactory accounts has yet been obtained from any of the sufferers. Whilst one has attributed this affliction to indulgence in spirituous liquors, and another to long lying on the damp ground; the others have been unable to suggest any circumstance whatever, which, in their opinion, could be considered as having given origin, or disposed, to the calamity under which they suffered.

Cases illustrative of the nature and cause of this malady are very rare. In the following case symptoms very similar are observable, so far as affecting the lower extremities. That the medulla spinalis was here affected, and in its lower part, is not to be doubted: but this, unfortunately, was never ascertained by examination. It must be however remarked, that this case differed from those which have been given of this disease, in the suddenness with which the symptoms appeared.

A.B. aged twenty-six years, during a course of mercury for a venereal affection, was exposed to severely inclement weather, for several hours, and the next morning, complained of extreme pain in the back, and of total inability to employ voluntarily the muscles of the lower extremities, which were continually agitated with severe convulsive motions. The physician who attended him employed those means which seemed best calculated to relieve him; but with no beneficial effect. The lower extremities were perpetually agitated with strong palpitatory motions, and, frequently, three or four times in a minute, suddenly raised with great vehemence two or three feet from the ground, either in a forward or oblique direction, striking one limb against the other, or against the chairs, tables, or any substance which stood in the way. To check these inordinate motions, no means were in the least effectual, except striking the thighs forcibly during the more violent convulsions. No advantage was derived from all the means which were employed during upwards of twelve months. Full ten years after this period, the unhappy subject of this malady was casually met in the street, shifting himself along, seated in a chair; the convulsive motions having ceased, and the limbs having become totally inert, and insensible to any impulse of the will.

It must be acknowledged, that in the well-known cases, described by Mr. Potts, of that kind of Palsy of the lower limbs which is frequently found to accompany a curvature of the spine, and in which a carious state of the vertebrae is found to exist, no instructive analogy is discoverable; slight convulsive motions may indeed happen in the disease proceeding from curvature of the spine; but palpitating motions of the limbs, such as belong to the disease here described, do not appear to have been hitherto noticed.

Whilst striving to determine the nature and origin of this disease, it becomes necessary to give the following particulars of an interesting case of Palsy occasioned by a fall, attended with uncommon symptoms, related by Dr. Maty, in the third volume of the Medical Observations and Inquiries. The subject of this case, the Count de Lordat, had the misfortune to be overturned from a pretty high and steep bank. His head pitched against the top of the coach, and was bent from left to right; his left shoulder, arm, and especially his hand, were considerably bruised. At first he felt a good deal of pain along the left side of his neck, but neither then, nor at any other time, had he any faintings, vomitings, or giddiness.—On the sixth day he was let blood, on account of the pain in his shoulder and the contusion of his hand, which were then the only symptoms he complained of, and of which he soon found himself relieved.—Towards the beginning of the following winter, he began to find *a small impediment in uttering some words, and his left arm appeared weaker.* In the following spring, having suffered considerably from the severities of the winter campaign, he found *the difficulty in speaking, and in moving his left arm, considerably increased.*—On employing the thermal waters of Bourbonne, his speech became freer, but, on his return to Paris, the Palsy was increased, and the arm somewhat wasted.—In the beginning of the next spring he went to Balaruc; when he became affected with *involuntary convulsive motions all over the body.* The left arm withered more and more, *a splitting began*, and now it was *with difficulty that he uttered a few words.* Frictions and sinapisms were successively tried, and an issue, made by a caustic, was kept open for some time without any effect; but no mention is made of what part the issue was established in.

Soon after this, and three years and a half after the fall, Doctor Maty first saw the patient, and gives the following description of his situation. "A more melancholy object I never beheld. The patient, naturally a handsome, middle-sized, sanguine man, of a cheerful disposition, and an active mind, appeared much emaciated, stooping, and dejected. *He still walked alone with a cane, from one room to the other, but with great difficulty, and in a tottering manner;* his left hand and arm were much reduced, and would hardly perform any motion; *the right was somewhat benumbed, and he could scarcely lift it up to his head; his saliva was continually trickling out of his mouth, and he had neither the power of retaining it, nor of spitting it out freely.* What words he still could utter were monosyllables, and these came out, after much struggle, in a violent expiration, and with such a low voice and indistinct articulation, as hardly to be understood but by those who were constantly with him. He fetched his breath rather hard; his pulse was low, but neither accelerated nor intermitting. He took very little nourishment, could chew and swallow no solids, and even found great pain in getting down liquids. Milk was almost his only food; his body was rather loose, his urine natural, his sleep good, his senses, and the powers of his mind, unimpaired; he was attentive to, and sensible of every thing which was said in conversation, and shewed himself very desirous of joining in it; but was continually checked by the impediment in his speech, and the difficulty which his hearers were put to. Happily for him he was able to read, and as capable as ever of writing, as he shewed me, by putting into my hands an account of his present situation, drawn up by himself: and I am informed that he spent his time to the very last, in writing upon some of the most abstruse subjects."

This gentleman died about four years after the accident, when the body was examined by Dr. Bellett and Mons. Sorbier, who made the following report:

"We first examined the muscles of the tongue, which were found extenuated and of a loose texture. We observed no signs of compression in the lingual and brachial nerves, as high as their exit from the basis of the cranium and the vertebrae of the neck; but they appeared to us more compact than they commonly are, being nearly tendinous. The dura mater was in a sound state, but the pia mater was full of blood and lymph; on it several hydatids, and towards the falx some marks of suppuration were observed. The ventricles were filled with water, and the plexus choroides was considerably enlarged, and stuffed with grumous blood. The cortical surface of the brain appeared much browner than usual, but neither the medullary part nor cerebellum were impaired. We chiefly took notice of the Medulla Oblongata, this was greatly enlarged, surpassing the usual size by more than one third. It was likewise more compact. The membranes, which, in their continuation, inclose the spinal marrow, were so tough that we found great difficulty in cutting through them, and we observed this to be the cause of the tendinous texture of the cervical nerves. The marrow itself had acquired such solidity as to elude the pressure of our fingers, it resisted as a callous body, and could not be bruised. This hardness was observed all along the vertebrae of the neck, but lessened by degrees, and was not near so considerable in the vertebrae of the thorax. Though the patient was but nine and thirty years old, the cartilages of the sternum were ossified, and required as much labour to cut them asunder as the ribs; like these they were spungy, but somewhat whiter. The lungs and heart were sound. At the bottom of the stomach appeared an inflammation, which increased as it extended to the intestines. The ileum looked of that dark and livid hue, which is observed in membranous parts tending to mortification. The colon was not above an inch in diameter, the rectum was smaller still, but both appeared sound.— From these appearances, we were at no loss to fix the cause of this gradual palsy in the alteration of the medulla spinalis and oblongata."

Dr. Bellett offers the following explanation of these changes. "I conceive, that, by this accident, the head being violently bent to the right, the nervous membranes on the left were excessively stretched and irritated; that this cause extended by degrees to the spinal marrow, which being thereby compressed, brought on the paralytic symptoms, not only of the left arm, but at last in some measure also of the right. This induration seems to have been occasioned by the constant afflux of the nutritive juices, which were stopt at that place, and deprived of their most liquid parts; the grosser ones being unable to spread in the boney cavity, by which they were confined, could only acquire a greater solidity, and change a soft body into a hard and nearly osseous mass. This likewise accounts for the increase of the medulla oblongata, which being loaded with more juices than it could send off, swelled in the same manner as the branches of trees, which will grow of a monstrous size, when the sap that runs into them is stopt in its progress. The medulla oblongata not growing so hard as the spinalis, was doubtless owing to its not being confined in an osseous theca, but surrounded with soft parts, which allowed it room to spread. The obstruction from the bulk of this substance must have affected the brain, and probably induced the thickening of the pia mater, the hydatids, and the beginning of suppuration, whereas the dura mater, being of a harder texture, was not injured.*"

In some of the symptoms which appeared in this case, an agreement is observable between it and those cases which are mentioned in the beginning of these pages. The weakened state of both arms; the power first lessening in one arm, and then in a similar manner in the other arm; the affection of the speech; the difficulty in chewing and in swallowing; as well as of retaining, or freely discharging, the spittle; the convulsive motions of the body; and the unimpaired state of the intellects; constitute such a degree of accordance as, although it may not mark an identity of disease, serves at least to show that nearly the same parts were the seat of the disease in both instances. Thus we attain something like confirmation of the supposed proximate cause, and of one of the assumed occasional causes.

Whilst conjecturing as to the cause of this disease, the following collected observations on the effects of injury to the medulla spinalis, by Sir Everard Home, become particularly deserving of attention. It thence appears, that none of the characteristic symptoms of this malady are produced by compression, laceration, or complete division of the medulla spinalis.

"Pressure upon the medulla spinalis of the neck, by coagulated blood, produced paralytic affections of the arms and legs; all the functions of the internal organs were carried on for thirty-five days, but the urine and stools passed involuntarily.**

"Blood extravasated in the central part of the medulla, in the neck, was attended with paralytic affection of the legs, but not of the arms.†

"In a case where the substance of the medulla was lacerated in the neck, there was a paralysis in all the parts below the laceration, the lining of the

---

*Medical Observations and Inquiries, Vol. III, p. 257.

**A coagulum of blood, the thickness of a crownpiece, was found lying upon the external surface of the dura-matral covering of the medulla spinalis, extending from the fourth vertebra colli to the second vertebra dorsi. The medulla spinalis itself was uninjured.

†The sixth and seventh vertebra colli were dislocated, the medulla spinalis, externally, was uninjured; but in the centre of its substance, just at that part, there was a coagulum of blood nearly two inches in length.

oesophagus was so sensible, that solids could not be swallowed, on account of the pain they occasioned.*

"When the medulla of the back was completely divided, there was momentary loss of sight, loss of memory for fifteen minutes, and permanent insensibility in all the lower parts of the body. The skin above the division of the spinal marrow perspired, that below did not. The wounded spinal marrow appeared to be extremely sensible."† *Philosophical Transactions*, 1816, p. 485.

In two of the cases already noticed, symptoms of rheumatism had previously existed; and in Case IV the right arm, in which the palpitation began, was said to have been very violently affected with rheumatic pain to the fingers ends. The consideration of this case, in which the palpitation had been preceded, at a considerable distance of time, by this painful affection of the arm, led to the supposition that this latter circumstance might be the cause of the palpitations, and the other subsequent symptoms of this disease. This supposition naturally occasioned the attention to be eagerly fixed on the following case; and of course influenced the mode of treatment which was adopted.

A.B. subject to rheumatic affection of the deltoid muscle, had felt the usual inconveniences from it for two or three days; but at night found the pain had extended down the arm, along the inside of the fore-arm, and on the sides of the fingers, in which a continual tingling was felt. The pain, without being extremely intense, was such as effectually to prevent sleep: and seemed to follow the course of the brachial nerve. Whilst ascertaining the propriety of this conclusion, the pain was found to ramify, as it were, on the fore and back part of the chest; and was slightly augmented by drawing a deep breath.

These circumstances suggested the probability of slight inflammation, or increased determination to the origin of the nerves of these parts, and to the neighbouring medulla. On this ground, blood was taken from the back part of the neck, by cupping; hot fomentations were applied for about the space of an hour, when the upper part of the back of the neck was covered with a blister, perspiration was freely induced by two or three small doses of antimonials, and the following morning the bowels were evacuated by an appropriate dose of calomel. On the following day the pains were much diminished, and in the course of four or five days were quite removed. The arm and hand felt now more than ordinarily heavy, and were evidently much weakened: aching, and feeling extremely wearied

*The seventh vertebra colli was fractured, and the medulla spinalis passing through it, was lacerated and compressed.

†The spinal marrow, within the canal of the sixth vertebra dorsi, was completely destroyed by a musket ball. The person lived four days.

after the least exertion. The strength of the arm was not completely recovered at the end of more than twelvemonths; and, after more than twice that time, exertion would excite the feeling of painful weariness, but no palpitation or other unpleasant symptom has occurred during the five or six years which have since passed.

The commencement, progress, and termination of this attack; with the success attending the mode of treatment, and the symptoms which followed, seem to lead to the conjecture, that the proximate cause of the disease, in this case, existed in the medulla spinalis, and that it might, if neglected, have gradually resolved itself into that disease which is the object of our present inquiry.

Some few months after the occurrence of the preceding case, the writer of these lines was called to a female about forty years of age, complaining of great pain in both the arms, extending from the shoulder to the finger ends. She stated, that she was attacked in the same manner as is described in the preceding case, about nine months before; that the complaint was considered as rheumatism, and was not benefited by any of the medicines which had been employed; but that after three or four weeks it gradually amended, leaving both the arms and hands in a very weakened and trembling state. From this state they were now somewhat recovered; but she was extremely anxious, fearing that if the present attack should not be soon checked, she might entirely lose the use of her hands and arms.

Instructed by the preceding case, similar means were here recommended. Leeches, stimulating fomentations, and a blister, which was made for sometime to yield a purulent discharge, were applied over the cervical vertebrae; and in the course of a very few days the pain was entirely removed. It is regretted that no farther information, as to the progress of this case, could be obtained.

On meeting with these two cases, it was thought that it might not be improbable that attacks of this kind, considered at the time merely as rheumatic affections, might lay the foundation of this lamentable disease, which might manifest itself at some distant period, when the circumstance in which it had originated, had, perhaps, almost escaped the memory. Indeed when it is considered that neither in the ordinary cases of Palsy of the lower extremities, proceeding from diseased spine, nor in cases of injured medulla from fractured vertebrae, any of the peculiar symptoms of this disease are observable, we necessarily doubt as to the probability of its being the direct effect of any sudden injury. But taking all circumstances into due consideration, particularly the very gradual manner in which the disease commences, and proceeds in its attacks; as well as the inability to ascribe its origin to any more obvious cause, we are led to seek for it in some slow morbid change in the structure of the medulla, or its investing membranes, or theca, occasioned by simple inflammation, or rheumatic or scrophulous affection.

It must be too obvious that the evidence adduced as to the nature of the proximate and occasional causes of this disease, is by no means conclusive. A reference to the test therefore which will be yielded by an examination of some of the more prominent symptoms, especially as to their agreement with the supposed proximate cause, is more particularly demanded. Satisfied as to the importance of this part of the present undertaking, no apology is offered for the extent to which the examination is carried on.

If the palpitation and the attendant weakness of the limbs, &c. be considered as to the order in which the several parts are attacked, it is believed, that some confirmation will be obtained of the opinion which has been just offered, respecting the cause, or at least the seat, of that change which may be considered as the proximate cause of this disease.

One of the arms, in all the cases which have been here mentioned, has been the part in which these symptoms have been first noticed; the legs, head, and trunk have then become gradually affected, and lastly, the muscles of the mouth and fauces have yielded to the morbid influence.

The arms, the parts first manifesting disordered action, of course direct us, whilst searching for the cause of these changes, to the brachial nerves. But finding the mischief extending to other parts, not supplied with these, but with other nerves derived from nearly the same part of the medulla spinalis, we are of course led to consider that portion of the medulla spinalis itself, from which these nerves are derived, as the part in which those changes have taken place, which constitute the proximate cause of this disease.

From the subsequent affection of the lower extremities, and from the failure of power in the muscles of the trunk, such a change in the substance of the medulla spinalis may be inferred, as shall have considerably interrupted, and interfered with, the extension of the nervous influence to those parts, whose nerves are derived from any portion of the medulla below the part which has undergone the diseased change.

The difficulty in supporting the trunk erect, as well as the propensity to the adopting of a hurried pace, is also referable to such a diminution of the nervous power in the extensor muscles of the head and trunk, as prevents them from performing the offices of maintaining the head and body in an erect position.

From the impediment to speech, the difficulty in mastication and swallowing, the inability to retain, or freely to eject, the Saliva, may with propriety be inferred an extension of the morbid change upwards through the medulla spinalis to the medulla oblongata, necessarily impairing the powers of the several nerves derived from that portion into which the morbid change may have reached. In the late occurrence of this set of symptoms, and the extension upwards of the diseased state, a very close agreement is observable between this disease and that which has been already shown, proved fatal to the Count de Lordat. But in this case, the disease doubtlessly became differently modified, and its symptoms considerably accelerated, in consequence of the magnitude of the injury by which the disease was induced.

Chapter V. Considerations Respecting the Means of Cure

The inquiries made in the preceding pages yield, it is to be much regretted, but little more than evidence of inference: nothing direct and satisfactory has been obtained. All that has been ventured to assume here, has been that the disease depends on a disordered state of that part of the medulla which is contained in the cervical vertebrae. But of what nature that morbid change is; and whether originating in the medulla itself, in its membranes, or in the containing theca, is, at present, the subject of doubt and conjecture. But although, at present, uninformed as to the precise nature of the disease, still it ought not to be considered as one against which there exists no countervailing remedy.

On the contrary, there appears to be sufficient reason for hoping that some remedial process may ere long be discovered, by which, at least, the progress of the disease may be stopped. It seldom happens that the agitation extends beyond the arms within the first two years; which period, therefore, if we were disposed to divide the disease into stages, might be said to comprise the first stage. In this period, it is very probable, that remedial means might be employed with success: and even, if unfortunately deferred to a later period, they might then arrest the farther progress of the disease, although the removing of the effects already produced, might be hardly to be expected.

From a review of the changes which had taken place in the case of Count de Lordat, it seems as if we were able to trace the order and mode in which the morbid changes may proceed in this disease. From any occasional cause, the thecal ligament, the membranes, or the medulla itself, may pass into the state of simple excitement or irritation, which may be gradually succeeded by such a local afflux and determination of blood into the minute vessels, as may terminate in actual but slow inflammation. The result of this would be a thickening of the theca, or membranes, and perhaps an increase in the volume of the medulla itself, which would gradually occasion such a degree of pressure against the sides of the unyielding canal, as must eventually intercept the influence of the brain upon the inferior portion of the medullary column, and upon the parts on which the nerves of this portion are disposed.

From this review, and assuming that the morbid changes in this disease may not be widely dissimilar from those which occurred in the case of Count de Lordat, the chance of relief from the proposed mode of treatment may appear to be sufficient to warrant its trial.

In such a case then, at whatever period of the disease it might be proposed to attempt the cure, blood should be first taken from the upper part of the neck, unless contra-indicated by any particular circumstance. After which vesicatories should be applied to the same part, and a purulent discharge obtained by appropriate use of the Sabine Liniment; having recourse to the application of a fresh blister, when from the diminution of the discharging surface, pus is not secreted in a sufficient quantity. Should the blisters be found too inconvenient, or a sufficient quantity of discharge not be obtained thereby, an issue of at least an inch and a half in length might be established on each side of the vertebral columna, in its superior part. These, it is presumed, would be best formed with caustic, and kept open with any proper substance.*

Could it have been imagined that such considerable benefit: indeed, that such astonishing cures, could have been effected by issues in cases of Palsy of the lower extremities from diseased spine? although satisfied with ascribing those cases to scrofulous action, we are in fact as little informed respecting the nature of the affection, inducing the carious state of the vertebrae, as we are respecting the peculiar change of structure which takes place in this disease. Equally uninformed are we also as to the peculiar kind of morbid action, which takes place in the ligaments of the joints; as well as that which takes place in different instances of deep seated pains and affections of the parts contained in the head, thorax, and abdomen, and in all which cases the inducing of a purulent discharge in their neighbourhood is so frequently productive of a cure. Experiment has not indeed been yet employed to prove, but analogy certainly warrants the hope, that similar advantages might be derived from the use of the means enumerated, in the present disease. It is obvious, that the chance of obtaining relief will depend in a great measure on the period at which the means are employed. As in every other disease, so here, the earlier the remedies are resorted to, the greater will be the probability of success. But in this disease there is one circumstance which demands particular attention; the long period to which it may be extended. One of its peculiar symptoms, Scelotyrbe festinans, may not occur until the disease has existed ten or twelve years, or more; hence, when looking for the period, within which our hopes of remedial aid is to be limited; we may, guided by the slow progress of the malady, extend it to a great length, when compared with that within which we should be obliged to confined ourselves in most other diseases.

---

*Cork, which has been hitherto neglected, appears to be very appropriate to this purpose. It possesses lightness, softness, elasticity and sufficient firmness; and is also capable of being readily fashioned to any convenient form. The form which it seems would be best adapted to the part, is that of an almond, or of the variety of bean called scarlet bean; but at least an inch and a half in length.

But it is much to be apprehended, as in many other cases, that the resolution of the patients will seldom be sufficient to enable them to persevere through the length of time which the proposed process will necessarily require. As slow as is the progress of the disease, so slow in all probability must be the period of the return to health. In most cases, especially in those in which the disease has been allowed to exist long unopposed, it may be found that all that art is capable of accomplishing, is that of checking its further progress. Nor will this be regarded as a trifle, when, by reference to the history of the disease, is seen the train of harassing evils which would be thus avoided.

But it seems as if there existed reason for hoping for more. For supposing change of structure to have taken place, it is extremely probable that this change may be merely increase in mass or volume by interstitial addition, the consequence of increased action in the minute vessels of the part. In that case, should the instituting of a purulent discharge, in a neighbouring part, act in the manner which we would presume it may—should it by keeping up a constant discharge, not merely alter the determination, but diminish the inordinate action of the vessels in the diseased part; and at the same time excite the absorbents to such increased action as may remove the added matter; there will exist strong ground for hope, that a happy, though slow restoration to health, may be obtained.

Until we are better informed respecting the nature of this disease, the employment of internal medicines is scarcely warrantable; unless analogy should point out some remedy the trial of which rational hope might authorize. Particular circumstances indeed must arise in different cases, in which the aid of medicine may be demanded: and the intelligent will never fail to avail themselves of any opportunity of making trial of the influence of mercury, which has in so many instances, manifested its power in correcting derangement of structure.

The weakened powers of the muscles in the affected parts is so prominent a symptom, as to be very liable to mislead the inattentive, who may regard the disease as a mere consequence of constitutional debility. If this notion be pursued, and tonic medicines, and highly nutritious diet be directed, no benefit is likely to be thus obtained; since the disease depends not on general weakness, but merely on the interruption of the flow of the nervous influence to the affected parts.

It is indeed much to be regretted that this malady is generally regarded by the sufferers in this point of view, so discouraging to the employment of remedial means. Seldom occurring before the age of fifty, and frequently yielding but little inconvenience for several months, it is generally considered as the irremediable diminution of the nervous influence, naturally resulting from declining life; and remedies therefore are seldom sought for.

Although unable to trace the connection by which a disordered state of the stomach and bowels may induce a morbid action in a part of the medulla spinalis, yet taught by the instruction of Mr. Abernethy, little hesitation need be employed before we determine on the probability of such occurrence. The power, possessed by sympathy, of inducing such disordered action in a distant part, and the probability of such disordered action producing derangement of structure, can hardly be denied. The following Case seems to prove, at least, that the mysterious sympathetic influence which so closely simulates the forms of other diseases, may induce such symptoms as would seem to menace the formation of a disease not unlike to that which we have been here treating of.

A.B. A man, 54 years of age, of temperate habits and regular state of bowels, became gradually affected with slight numbness and prickling, with a feeling of weakness in both arms, accompanied by a sense of fulness about the shoulders, as if produced by the pressure of a strong ligature; and at times a slight trembling of the hands. During the night, the fullness, numbness, and prickling were much increased. The appetite had been diminished for several weeks; and the abdomen, on being examined, felt as though containing considerable accumulation.

Before adopting any other measures, and as there appeared to be no marks of vascular fullness, it was determined to empty the bowels. This was done effectually by moderate doses of calomel, with the occasional help of Epsom salts; and in about ten days, by these means alone, the complaints were entirely removed.

Before concluding these pages, it may be proper to observe once more, than an important object proposed to be obtained by them is, the leading of the attention of those who humanely employ anatomical examination in detecting the causes and nature of diseases, particularly to this malady. By their benevolent labours its real nature may be ascertained, and appropriate modes of relief, or even of cure, pointed out.

To such researches the healing art is already much indebted for the enlargement of its powers of lessening the evils of suffering humanity. Little is the public aware of the obligations it owes to those who, led by professional ardour, and the dictates of duty, have devoted themselves to these pursuits, under circumstances most unpleasant and forbidding. Every person of consideration and feeling, may judge of the advantages yielded by the philanthropic exertions of a Howard; but how few can estimate the benefits bestowed on mankind, by the labours of a Morgagni, Hunter, or Baillie.

FINIS.

Parkinsonism Since Parkinson

F. Clifford Rose

The history of Parkinson's disease has been divided into six periods (Hesselink, 1983), the first being from Galen until 1817, when Parkinson's *Essay* was published; the second from 1817 until Charcot's 1868 paper, during which period there was still confusion between Parkinson's disease, chorea, and multiple sclerosis; the third from 1868 until the end of the century, when the distinctions between these disorders became clear and the lesion was localised to the substantia nigra (Blocq and Marinesco, 1893; Brissaud, 1895); the fourth from the beginning of this century until Kinnier Wilson's lectures of 1925, a period which included the dissertation of Tretiakoff (1919) confirming Brissaud's hypothesis; the fifth was until the 1950s, when the dopamine story began; and the sixth and final period lasts until the present.

From Parkinson to Charcot (1817–1868)

Though others had recognized the contribution of James Parkinson, it was Charcot who, 37 years after the founder's death, called it Parkinson's disease (Charcot and Vulpian, 1861).

Although Bechet (1892) stated that Parkinson's disease remained almost unknown until 1850, there had been several mentions of it, as well as case reports. In what has been claimed the first comprehensive treatise on nervous disease (Schiller, 1986), Cooke (1821) had taken note of the *Essay*: "Mr. Parkinson's *paralysis agitans* or shaking palsy . . . appears to me to be highly deserving of our attention . . ." and pointed out that "the limbs, in walking, cannot be raised to that height, nor with that promptitude, which the will directs," i.e. bradykinesia. Probably the first case report after Parkinson's *Essay* was that of Toulmouche (1833), a 76-year-old man who had developed tremor in his legs nine years previous and whose walking, a year later, had become slowed and with a tendency to loss of balance. The involuntary movements spread to his arms and his intellectual capacity became diminished. Although Toulmouche made no reference to Parkinson, there was no doubt about the diagnosis (Charcot and Vulpian 1861).

Six years later Elliotson (1839) quoted from the *Essay* and reported a case with head tremor, a sign that Charcot (1868) denied as a feature of the disorder. Elliotson distinguished two types, the first occurring in young people which could be cured, while the second was due to an organic "induration process" within the central nervous system and was incurable. This subdivision could be the result of the diagnostic confusion between Parkinson's disease and multiple sclerosis, his young patients probably belonging to the latter group.

After Elliotson, several others described cases of Parkinson's disease, including Marshall Hall (1841) who, like Parkinson, thought that the lesion was in the medulla. He recognized that the disease began unilaterally (i.e. hemiparkinsonism) and could spread to the other side within a year. Because his patient was only 28 years old and had nystagmus—"a peculiar rolling motion of the eyes"—his case of "hemiplegic parkinsonism" was most likely to be multiple sclerosis. Graves (1842) described Elliotson's report as an "excellent description" and was also fascinated by the characteristic gait. While these papers added hemiparkinsonism and retropulsion to the phenomenology of Parkinson's disease, confusion with multiple sclerosis, idiopathic tremor, and symptomatic parkinsonism remained.

Even more than 40 years after the *Essay*, Parkinson's disease was considered rare; i.e. Cohn (1860) wrote, "The literature on this subject is extremely small. . . ." This may be because life expectation was much less a century ago, and Parkinson's disease is an age-related disorder. Three decades later, Parkinson's disease was by no means uncommon (Osler, 1892).

Tremor

Copland (1858) indicated that the tremor could start in the head and would increase with emotion or exercise. While Parkinson had emphasized that the tremor was at rest, Cohn (1860) reported other types, one of which was present on intention, and attempted to correlate these clinical findings with those found at autopsy. It was the three articles of Charcot and Vulpian (1861–1862) which distinguished parkinsonian from other tremors. Handfield Jones (1864) also distinguished between a juvenile type (probably multiple sclerosis) and that occurring in older age groups, but it was Charcot four years later who gave the more precise delineation, namely that the tremor of multiple sclerosis occurred on movement whereas that of Parkinson's disease was present at rest. Hughling Jackson (1899) thought that parkinsonian tremor was caused by damage to the cerebral hemisphere given a release phenomenon due to the unopposed action of the cerebellum; rigidity, he thought, was a confluent tremor.

Rigidity

As with Trousseau (1867) before him, Charcot (1868) pointed out that Parkinson had not reported muscle rigidity in spite of noting the flexed posture of the body due to the differential increase of tone in spinal flexors and extensors. When Parkinson wrote his *Essay*, the assessment of tone was not yet part of the clinical examination, since it was only in the middle of the last century that the examination of the nervous system moved from simple observation to specific feats, e.g. Romberg's sign (1851).

The word, tone, derives from the Greek *tonos* from the verb—*tenein*—to stretch (as does the word "tone" in music, i.e. from stretching a string and "tonic", i.e. agents combatting a letdown) (Schiller, 1986). At that time it was considered that all muscles "are constantly ever in the state of rest, subject to the influence of the nervous principle" (Muller, 1842), a view not disproved until the advent of electromyography more than a century later (Ralston and Liber, 1953).

Von Oppolzer (1861) was the first to report a case of paralysis agitans where "the affected muscles are contracted." Later, Trousseau also reported a case of paralysis agitans where there was a "permanent state of contraction, and the sensation complained of that of a continuous effort . . . He sat down with some difficulty as if his trunk and legs were stiff."

Von Oppolzer's report stimulated the detailed account by Charcot and Vulpian (1861–1862), who described "the contractures as the sensation of stiffness which cannot be overcome as it appears in the limbs affected by the tremor . . . Yet this (debility) is so pronounced that the patients become incapacitated for getting out of bed, or even to perform any major movement . . . later usually their mental functions markedly decrease."

It was Benedikt (1868), a pupil of Von Oppolzer, who first reported the increase in patients' resistance "most frequently with passive movements: visibly and palpably muscles become tense in their entirety . . . Usually the muscles that are being stretched are so affected, rarely the shortened ones, . . . The contraction against Stretching is felt throughout, or only at certain moments. . . ." Sinkler (1886) was the first to point out the "absolute absence of expression."

In Gower's classical textbook (1888) he reports that the "muscular weakness and rigidity usually come together . . . voluntary movement is not only feeble, it is also slow. This seems to be in part at least the result of muscular rigidity, which causes a resistance to passive movements . . . (which) preponderates in the flexors." Careful testing of tone in parkinsonism showed that it was not smooth but regularly interrupted, and this was first called "cogwheel" resistance by Moyer (1911).

Akinesia

Charcot (1877) emphasized rigidity and bradykinesia—"Laboriousness in the execution of movements, dependent neither on the existence of tremors, nor on that of muscular rigidity . . . There is retardation in the execution of movements rather than any real enfeeblement of the motor powers . . . They are able to accomplish most to the motor acts . . . but . . . with extreme slowness." Charcot also indicated the difficulty in initiating movements and getting out of a chair (i.e. akinesia). Akinesia can be either narrowly defined as rigid posture, diminished arm swing and short-stepped gait or more widely to include lack of speed, spontaneity, emotional reactivity and interest in work. Later (1880), he pointed out the characteristic posture of the body and hands, and micrographia.

Following Charcot's seminal observations, several personal series were published, including that of Collins and Muskins (1889), Gower (1893), Erb (1901) and the largest, of 219 cases, by Hart (1904). While Charcot recognized three phases in the disease: an initial mild phase, the fully developed disease, and complete invalidism, Saunders (1859) recognized five stages: unilateral symptoms, bilateral manifestations, falling while walking, inability to walk, and cachexia—not dissimilar from more modern classifications (Hoehn and Yahr 1967).

Site of the Lesion

Parkinson thought that the site of the lesion was in the medulla, a view that was still getting support nearly a quarter of a century later (Hall, 1841). Although glial cells were first delineated by Virchow in 1846, it was only when histological staining techniques developed in the second half of the 19th century that the pathological basis of brain disease became accurate. In spite of this, even as late as 1882, Berger suggested that trauma to peripheral nerves, emotional upset, or excessive cold could produce Parkinson's disease. Charcot thought the disease was due to neurosis and attempted treatment with psychotherapy.

Gowers (1888) thought the tremor was due to diffuse cerebral disease involving cortex, thalamus and internal capsule, while Hughlings Jackson (1899) thought the cause lay in the cerebellum.

Hammond (1871) thought that "One probable seat of the morbid process is the corpus striatum." Charcot's student, Oldenstein (1887), analyzed 30 of his teacher's cases of Parkinson's disease, two of which had been autopsied, and one of these showed "maceration" of the substantia nigra.

It was Blocq and Marinesco's (1893) case of a tuberculoma involving the substantia nigra that provided further evidence of the involvement of this nucleus. It was on this case that Brissaud (1895) based his hypothesis of the involvement of the substantia nigra, deducing on clinical grounds that "the localization of Parkinson's disease must be sub-thalamic or peduncular." Lewy (1914) had published a paper based on the examination of 60 brains from patients with Parkinson's disease. In the nucleus basalis of Meynert and the dorsal motor nucleus of the vagus he found inclusion bodies in the nerve cells which were rounded or worm-shaped and stained with eosin.

At a meeting in Paris on Parkinson's disease, Tretiakoff suggested that lesions in the substantia nigra or lentiform nucleus might be responsible, and in his thesis for the University of Paris (1919) he reported his examination of the substantia nigra (locus niger de Soemmering) in different neurological diseases. He found loss of pigmented cells in all nine patients with Parkinson's disease, six of whom had Lewy bodies. Because of the reports of Ramsay Hunt, the Vogts, and Lhermitte and Cornil suggesting other sites, the substantia nigra was not confirmed until Hassler (1938) carried out the most detailed study and proved the work of Tretiakoff. Because of the World War II, Hassler's contribution was not fully recognized. After the war, the controversy continued with Denny-Brown (1962) playing down the importance of the substantia nigra and indicating the loss of cortico-striatal fibers. Purdon Martin (1967) suggested the globus pallidus as the site of the lesion even though the consensus of neuropathologists at that time was that the "indispensable change" was loss of cells from the substantia nigra (Earle, 1963). It was also agreed that Lewy bodies, but not arteriosclerotic changes, were associated with Parkinson's disease. In 1918, von Economo showed that the substantia nigra was involved in encephalitis lethargica, a condition that could produce a clinical picture like Parkinson's disease.

Other Basal Ganglia

It was Meynert in 1868 who first postulated a relationship between involuntary choreiform movements and the basal ganglia and attempted to map out these central pathways by serial sections in 1883. By 1896, Anton considered that the causative lesion of Huntington's disease lay in the basal ganglia, but it was not until 1911 that the pathological basis of Huntington's disease, including the marked loss of ganglion cells in the corpus striatum, was described by Alzheimer, even though the clinical features of the condition had been known for decades. It was Ramsay Hunt (1916, 1917) who first thought that the dementia was due to cortical changes.

Up to the end of the 19th century, there was still no universal agreement as to the nomenclature or function of the basal ganglia; that involuntary movements were due to lesions of the basal ganglia rather than the cerebral cortex gradually became accepted. The interrelationship of the basal ganglia was first identified by Spatz, in 1927, who included in the basal ganglia the corpus striatium, pallidum, corpus Luysii (subthalamic nucleus), substantia nigra, red nucleus and dentate nucleus. Although the first description of a thalamic syndrome was by Dejerine in 1890, the full anatomical relationships of the thalamus had to wait until the work of Hassler. The term "extrapyramidal system" was first used by Prus (1899) but Kinnier Wilson conceived the idea of the basal ganglia coordinating muscle tone and voluntary movement, a concept not fully established until well into this century.

Post-encephalitic Parkinsonism

The Registrar-General (of the U.K.) has recorded annual deaths due to the Shaking Palsy from 1855 and as paralysis agitans from 1880. These figures indicated that the disease has not changed in prevalence (Duvoisin and Schweitzer, 1966). During the World War I, a "new" disease, encephalitis lethargica, became known, spreading from central Europe in 1915 to the rest of the world and ending in 1926. The mortality rate was more than 40% but half the survivors developed parkinsonism within five years; this was an underestimate of those who had minimal signs, e.g. facial hypomimia. In *the Lancet* of April 20, 1918 there were two case reports, one by Wilfred Harris, under the title "Acute infective ophthalmoplegia, or Botulism" and the other entitled "Note on an Epidemic of Toxic Opthalmoplegia Associated with Acute Strenic and Other Nervous Manifestations" (Hall, 1918). In the following issue, four letters appeared on similar cases, making a total of 21 cases, one of which was from Farquar Buzzard who reported a man aged 45 with a history of tiredness and vomiting for less than one week; examination showed "every aspect of paralysis agitans." These were the first British cases of the epidemic of encephalitis lethargica reported the previous year by Von Economo in Vienna. There had been previous epidemics in Europe in 1580, in London in 1674 (described by Sydenham as *"febris comatosa"*), in Germany in 1712, and in Italy (where it was called "nona") in 1889. It was the description of nona given to Von Economo that allowed him to recognize the 1917 epidemic (Sachs 1971). Von Economo (1920) clearly distinguished idiopathic Parkinson's from the post-encephalitic syndrome, and it was these latter cases that showed that parkinsonism need not be idiopathic and that there is more than one aetiology. A similar condition sometimes followed by parkinsonism had been reported 30 years previously. Von Economo (1931) did not think he was the first to describe the disorder. "We may assume with some degree of certainty that encephalitis lethargica had already appeared repeatedly before the Great War, both sporadically in the shape of isolated cases and in epidemics which again and again attracted notice for a short period on account of the singular combinations of symptoms displayed. These symptoms included fever with somnolence, slow movements, rigidity with a mask-like facies and tremors as well as oculogyric crises and dystonias."

Drug-Induced Parkinsonism

Soon after its introduction in 1952, chlorpromazine, the first neuroleptic, was found to produce symptoms similar to Parkinson's disease. Delay et al. (1952) coined the term neuroleptic, i.e. a drug producing a diminution of neurological activity with control of excitement and motor slowing. Patients also developed a mask-like facies and other parkinsonian features but the development of extrapyramidal side effects did not correlate with psychotic improvement. The restlessness produced by neuroleptics—akathisia—was difficult to distinguish from anxiety only if other extrapyramidal features were not present.

Therapy

Parkinson stated that "until we are better informed respecting the nature of this disease the employment of internal medicines is scarcely warrantable." This did not stop a whole host of remedies being tried. While Parkinson recommended blood letting, others tried iodides, phosphorus, radiotherapy, heavy metals, and even arsenic. Charcot (1880) was aware that hyoscine improved Parkinsonian tremor and, over the next 80 years, anticholinergic drugs were the only effective remedy. (This misled researchers into thinking that a failure in cholinergic brain mechanisms was the cause of Parkinson's disease).

Dopamine

It was Von Euler who first recognized that the post-ganglionic sympathetic neurotransmitter was norepinephrine. Since dopamine was an obligatory intermediate in the biosynthesis of norepinephrine (as well as that of epinephrine), it began to assume greater significance.

Rapid advances occurred only with the discovery of mono-amine neurotransmitters which were shown to have specific areas of concentration, e.g. noradrenaline (Vogt, 1954) and dopamine (Carlsson, 1959) and be localised *in situ* (Falck et al., 1962). It was the invention of formaldehyde-induced fluorescence (using vapourized formaldehyde to condense primary and secondary monamines to produce fluorogenic substances) that made it possible to identify such monamine neurotransmitters as dopamine, norepinephrine, and serotonin within the brain (Falck et al. 1962).

Montagu (1952) first found dopamine in mammalian brains, and it was in 1955 that it was first demonstrated that reserpine could reduce amine levels in the brain (Pletscher et al., 1955). Reserpine, the active ingredient of an ancient Indian herbal remedy, Rauwolfia serpentina, used for many years as an anti-psychotic and anti-hypertensive, had, as one of many side-effects, a parkinsonian syndrome. Four years later Carlsson found that extrapyramidal symptoms induced in rabbits by reserpine was due to reduced dopamine in the corpus striatum, and these symptoms could be reversed by dopa, the precursor of dopamine (Carlsson, 1959). It was he who first suggested giving levodopa for Parkinson's disease. Two years later, Carlsson found that dopamine was concentrated in the neostriatum and suggested it was linked with the extrapyramidal motor system.

In 1960 in the Department of Pharmacology of the University of Vienna, the changes in dopamine and noradrenaline in the brains of 40 parkinsonian patients were being compared with controls. Not only was there a decrease in dopamine in the basal ganglia of parkinsonian patients but there was a decrease in its metabolite, homovanillic acid, and the synthesising enzymes, tyrosine hydroxylase and L-dopa decarboxylase. Although L-dopa (levodopa, L-3, 4-dihydroxyphenylalanine) was synthesized early in this century and soon afterwards isolated from plants, it was not until 1938 that an enzyme that converted levodopa into dopamine (3-hydroxytyramine) was discovered and named dopa decarboxylase. The degree of dopamine deficiency correlated well with the degree of degeneration of the pars compacta of the substantia nigra. All these findings confirmed that dopamine was depleted in parkinsonian brains (Ehringer and Horneykiewicz 1960). In the following year it was discovered that there were lower levels of dopamine in the urine of parkinsonian patients (Barbeau et al, 1961).

The first successful treatment was reported by Birkmeyer and Horneykiewicz (1961), who gave intravenous levodopa in doses of 50–150 mg and were able to improve motor performance and gait, the effect reaching its peak in 2–3 hours and disappearing within 24 hours. Over the next three years the same authors reported their experience in 200 parkinsonian patients (Birkmeyer and Horneykiewicz, 1964).

Barbeau reported the preliminary results of the first clinical trial of oral L-dopa in Parkinson's disease at the 7th International Congress of Neurology in Rome in 1961. With two grams of levodopa he had noticed definite improvement of both rigidity and akinesia within one-half hour, but disappearing within two and a half hours (Barbeau et al., 1962). These findings provoked investigations into the dopaminergic pathway with findings that had anatomical, physiological, and biochemical implications. The melanin-containing cells of the substantia nigra were found to terminate in the caudate nucleus and putamen, dopamine was clearly a neurotransmitter, and its lack played a role in akinesia and rigidity. Although the dopaminergic nigro-striatal pathway was worked out in animals between 1964 and 1966 this followed the initial attempts to treat patients with levodopa.

The full recognition of the value of levodopa was not generally accepted until several years later, when much larger doses were used which saturated the enzymes broke down dopa to dopamine, Cotzias et al. (1967) used dl-dopa in a dose of 3–16 g per day to obtain very good results, but four of their patients developed blood dyscrasias. In the following year they reported that L-dopa gave the same good results but without the side effects (Cotzias, 1968, Cotzias et al., 1968).

Later, Barbeau recognized that the three main problems of L-dopa therapy work were the wearing off effect (end-of-dose akinesia), the on-off phenomenon, and the freezing phenomenon (akinesia paradoxica) (Barbeau, 1981).

Amantadine

The usefulness of this drug in Parkinson's disease was discovered serendipitously. Schwab et al. (1969) reported "In early April, 1968, a 58-year-old woman with moderately severe bilateral Parkinson's disease recounted to us that 3 months before, while taking amantadine hydrochloride 100 mg twice daily to prevent the flu, she experienced a remarkable remission in her symptoms of rigidity, tremor and akinesia. These promptly returned when stopping the drug after 6 weeks." The effects of amantadine are midway between the anticholinergics and levodopa.

Decarboxylase Inhibitors

Because dl-dopa caused agranulocytosis, albeit reversible, the levoisomer was then used. Its most troublesome side effect was nausea and vomiting, which was due to dopamine stimulating the vomiting centre, i.e. the area postrema of the medulla, a chemoreceptor trigger zone outside the blood-brain barrier. Other side effects were postural hypotension and cardiac dysrrhythmia. In order to diminish these, inhibitors of the enzyme (aromatic acid decarboxylase) that converts levodopa to dopamine were given, either carbidopa (Porter et al., 1962) or benzerazide (Bartholini et al., 1967). This enabled the amount of levodopa given to be reduced by more than half.

Dopamine Agonists

Because of the problems with levodopa, attempts have been made to stimulate dopamine receptors directly. Experimentally it was shown that an ergot alkaloid, bromocriptine, could stimulate receptors in the animal model (Corrodi et al., 1971) and later it was shown to work in humans by Calne et al. (1974).

Selegeline

Because the enzyme monoamine oxidase (MAO) breaks down dopamine, MAO inhibitors have been tried. Birkmeyer and his colleagues (1975) showed that giving selegeline, an MAO B inhibitor, could reduce the functional deficit within an hour and the effect could last up to three days. They confirmed its long term benefit two years later, and produced some evidence that the lives of parkinsonian patients could be prolonged by its use.

Lesioning the Motor Pathway

Parkinson had reported the relief of tremor following a hemiplegia and its return with the recovery of motor power.

The relief of hemi-athetosis by cortical resection was reported by Sir Victor Horsley (1909). Bucy tried this technique for parkinsonism, but this resulted in a 20% post-operative mortality with marked hemiplegia and often convulsions.

Because of the latter complications, Putnam (1938) sectioned the crossed pyramidal tract at the level of the second cervical segment of the spinal cord; of his 22 cases, 15 showed an improvement in tremor, greatest in those with the most marked paralysis. When motor power returned, so did the tremor. Later surgeons tried sectioning both crossed and uncrossed pyramidal tracts (Ebin, 1949), the lateral columns (Oliver, 1950) and the cerebral peduncle (Walker, 1952).

Other Surgical Techniques

Because of the inadequacies of medical therapy, a whole variety of surgical therapies have been tried. For rigidity, posterior rhizotomy was done in the early part of this century as well as sympathectomy either by excising the ganglia or sectioning the rami. For tremor, sectioning of the rubrospinal tract or the dorsal column was tried.

Kinnier Wilson during the 1920s thought that producing lesions in corpus striatum of animals would give a similar symptomatology of the disease named for him. He used to stereotactic machine devised by Horsley and Clarke (1920), but these workers became estranged and would not speak to each other. The result was that each had one-half of the instrument and no one could use it (Schaltenbrand, 1975). The machine was faithfully copied and used on experimental animals at the Johns Hopkins Hospital in 1928. The first stereotactic lesions in man were performed in 1947 by Spiegel and Wycis and later by Narabayashi in Japan.

The next advance was when, during pedunculotomy for parkinsonism, Irving Cooper (1952) accidentally damaged one of the cerebral arteries and had to clip the choroidal artery, which produced cure of tremor and rigidity without loss of power or sensation. This was later found to be due to infarction of the substantia nigra and the medial part of the globus pallidus. Stereotactic lesions were made at different sites but the most effective for tremor was destruction of the ventrolateral nucleus of the thalamas (Reichert, 1958).

The MPTP Model

Another advance was the discovery that MPTP was a toxin that could induce parkinsonism (Langston et al., 1983). This was first noticed in intravenous drug users in California. Not only has this provided further evidence in support of a possible environmental cause for the condition, but it has also provided an animal model for the first time. The MPTP animal model differs from idiopathic parkinsonism pathologically in that, although eosinophilic inclusion bodies have been described, typical Lewy bodies are not usually found and the dopaminergic neuronal loss is less widespread; there are also clinical differences with spontaneous recovery occurring in the non-primate model; in spite of these differences, the model provides the opportunity for drugs to be tested both for their therapeutic effect and their protective effect on the substantia nigra. Other promising areas are those of imaging and transplantation (Rose and Nashef, 1987).

Neural Transplant

Although the idea of neural transplantation was initially for investigating development and regeneration in the central nervous system, it was found that grafting of cells, particularly foetal ones, could help neurological deficit. The first attempt was in 1890 of the cat cerebral cortex to dogs (Thompson, 1890). Autografts of cortical neurones survived in young rabbits for over one week (Salsykow, 1905). There was also successful grafting of neonatal rat spinal ganglia into adult rat cortex (Ranson, (1909). Cortical neurones from rat embryos survived if juxtaposed to the choroidal plexus of the lateral ventricles (Dunn, 1917), indicating the importance of vascularisation. There seemed little doubt that the greatest potential lay with fetal grafts (LeGros Clark, 1940), particularly in relationship to the ventricular system (Flerko and Szentágothai, 1957). The ethical aspects of foetal cell grafting is problematical and continues to provide the basis for much discussion. The operation has been done in over 100 patients since then but reports of improvement have been far from unanimous (Lindvall et al., 1987).

The first clinical trials using adrenal medullary transplant into the striatum were performed in Sweden (Backlund et al., 1985) when unilateral autografts were given to two parkinsonians: a 55 year old patient with an eight-year history of parkinsonism and a 46 year old woman with a five-year history. Tremor and hypokinesia were absent until the fourth postoperative day. Although the little clinical improvement did not persist, two years later, a group from Mexico City (Madrazo et al., 1987) reported a series of eight patients where adrenal medullary autografts were placed in the caudate nucleus and the adrenal fragments exposed to the CSF. Three of these patients were under the age of 50 and improved markedly for at least 10 months (the length of the study) but the five patients over 50 did not improve as much (Sladek and Gash, 1988).

These are still early days and as yet no consensus has been agreed on the value of these operations which are now being done in many centres. There is little doubt that James Parkinson would have been thrilled and excited by the increasing knowledge accrued since he first wrote his *Essay* describing six patients with the disorder named for him.

References

Abraham, James Johnston (1933). *Lettsom, His Life, Friends and Descendants.* London, W. Heinemann Medical Books.

Aita, J.F. and O'Connell, B.L. (1987). Maladie de Parkinson unilateral: symptomatique d'une tumeur de pedoncle cerebral. *Neurology* 37:1145–1146.

Alexander, J.H. (1887). *Letter to the British Medical Journal.* Vol. I, p. 1085.

Althaus, Julius (1877). *Diseases of the Nervous System, Their Prevalence and Pathology.* London, Smith, Elder and Co.

Anon. (1607). *The Pleasant Walks of Moore Fields, A Dialogue Between a Country Gentleman and a Citizen.* London, N.P.

Anon. (1797). *The Trial of John Smith . . . on December 6, 1796, for selling a work entitled: A summary of the duties of citizenship, London.* (In the Wellcome Institute for the History of Medicine, London.)

Association for Preserving Liberty & Property Against Republicans & Levellers (1792). *Address to the Public, Together with a Resolution Relative to Disloyal Newspapers. . . .* London, N.P.

Backlund, E.O., Granberg, P.O., Hamberger, B., Knutsson, E., Martensson, A., Sedvall, G., Sieger, A., and Olson, L. (1985). Transplantation of adrenal medullary tissue to striatum in parkinsonism. First clinical trials. *J. Neurosurg.* 62:169–173.

Bampton, A.H. (1885). *Letter to the Lancet.* Vol. II, p. 189.

Barbeau, A. (1981). The L-dopa Story. In: *Research Progress in Parkinson's Disease* (F. Clifford Rose and R. Capildeo, eds.). London, Pitman, pp. 221–225.

Barbeau, A., Murphy, B.F., and Sourkes, T.L. (1961). Excretion of Dopamine in diseases of the basal ganglia. *Science* 133:1706–1707.

Bartholini, G., Burkard, W.P., Pletscher, A., and Bates, H.M. (1967). Incidence of cerebral catecholamines caused by 3,4-dihydrosyphenylalamine after inhibition of peripheral decarboxylase. *Nature* 215:852–853.

Bechet, E. (1892). Former cliniques et diagnostic de la maladie de Parkinson. Thèse, Paris.

Benedikt, M. (1868). *Elektrotherapie.* Wien, Tendler, pp. 134, 149–150.

Besant, Walter (1886). *Children of Gibeon.* London, Chatto & Windus.

Besant, Walter (1911). *London in the Eighteenth Century.* London, Chatto & Windus.

Bingham, Frederick (1921). *The Official Guide to the Metropolitan Burrough of Shoreditch.* London, N.P.

Birkmeyer, W. and Horneykiewicz, O. (1961). Der L-3, 4-dioxyphenylalin (= DOPA) Effekt bei der Parkinson akinese. *Wein Klin. Wschr.* 73:787–788.

Birkmeyer, W. and Horneykiewicz, O. (1964). Weitere Experimentale Untersuchungen uber L-dopa beim Parkinson-syndrom und reserpine-parkinsonismus. *Arch. Psychiat. Nervenkr.* 206:367–381.

Blocq, P. and Marinesco, G. (1893). Sur un cas de tremblement parkinsonian hemiplegique symptomatique du a une tumour du pedoncle cerebral. *Rev. Neurol.* 2:265.

Booth, Charles (1902). *Life and Labour of the People in London.* 17 vols., London, Macmillan & Co.

Bradley, H.J. (1914). *The History of Shoreditch Church.* London, N.P.

Brissaud, E. (1895). Clinique des maladies du systeme nerveux: sur la nature et la pathogenie de la maladie de Parkinson. *Journal de Medicine et la Chirurgie Pratiques* 65:641–663.

Bynum, W.B. (1989). *Personal communication.* Wellcome Institute for the History of Medicine, London.

British Medical Journal (1888). *Pathogeny of Paralysis Agitans.* Vol. II, p. 130.

Buchan, William (1769). *Domestic Medicine; Or, The Family Physician.* Edinburgh, Belfour, Auld & Smellie.

Burdett, Sir Francis (1809). *Parliamentary Reform. A Full & Accurate Report of the Proceedings at the Meeting Held at the Crown and Anchor Tavern, on Monday 1st May 1809. . . .* London, N.P.

Burdett, Sir Francis (1810). *The Address of Sir F. Burdett Dated March 23 to his Constituents, in a Letter, Denying the Power of the House of Commons to Imprison the People of England.* London, T. Broom.

Burgh, James (1774–7). *Political Disquisitions: Or, An Enquiry into Public Errors, Defects, and Abuses. . . .* London, E. & C. Dilly.

Burke, Edmund (1770). *Thoughts on the Cause of the Present Discontents.* London, J. Dodsley.

Burke, Edmund (1778). *A Vindication of Natural Society; Or, A View of the Miseries and Evils.* London, R. & J. Dodsley.

Burke, Edmund (1790). *Reflections on the Revolution in France & On the Proceedings of Certain Societies in London Relative to that Event.* London, J. Dodsley.

Burke, Edmund (1796). *Thoughts on the Prospect of a Regicide Peace. In a Series of Letters.* London, J. Owen.

Buzzard, Thomas (1882). *Clinical Lectures on Diseases of the Nervous System.* London, J. & S. Churchill.

Byfield, Timothy (1687). *A Short and Plain Account of the Balsamic Wells of Hoxdon.* London, Christopher Wilkinson.

Calne, D.B., Teychenne, P.F., Claveria, L.E., Eastman, R., Greenacre, and Petrie, A. (1974). Bromocriptine in parkinsonism. *Brit. Med. J.* 4:442–444.

Carlsson, A., Lindquist, M., and Magnusson, T. (1953). 3,4-dihydroxyphenylalanine and 5-hydroxythyptophan as reserpine antagonists. *Nature* 180:1200.

Carlsson, A. (1959). Detection and assay of dopamine. *Pharmacol. Rev.* 11:300–304.

Carlsson, A. (1959) The occurrence, distribution and physiological role of catecholamines in the nervous system. *Pharmacol. Rev.* 11:490–493.

Charcot, J.M. (1877). *Lectures on Diseases of the Nervous System.* Lecture V, p. 129. London, The New Sydenham Society.

Charcot, J.M. (1877). On paralysis agitans. In: *Lectures on the Diseases of the Nervous System* (translated by G. Sigerson). London, New Sydenham Society, pp. 129–156.
Charcot, J.M. (1880). *Lecons sur les maladies du systeme nerveux faites a la Salpetriere* (4th edition, collected and published by Bourneville). Dalahaye and Lecrosnier, p. 186.
Charcot, J.M. and Vulpian, E.F. (1861–62). De la paralysie agitante. (A propos d'un cas tiré de la Clinique du Professeur Oppolzer.) *Gaz. Hebdom. Med. Chirur.* 8:765–767, 816–860; and 9:54–59.
Clark-Kennedy, A.E. (1962–1963). *The London: A Study of the Voluntary Hospital System.* London, Pitman Medical.
Cohn, B. (1860). Ein Beitrag zur Lehre der Paralysis Agitans. *Wien Med. Wehschn.* 10:272–277.
Collins, J. and Muskens, L.J.J. (1889). A clinical study of twenty-four cases of paralysis agitans. *N.Y. Med. J.* 70:41–46.
Colquhoun, Patrick (1806). *A Treatise on Indigence; Exhibiting a General View of the National Resources for Productive Labour.* London, J. Hatchard.
Conybeare, W.D. and Phillips, W. (1822). *Outlines of the Geology of England and Wales.* London, W. Phillips.
Cooke, J. (1821). *A Treatise on Nervous Disorders. Vol. 2.* London, Longman, Hurst, Rees, Orme, and Brown, pp. 207–210.
Copland, J. (1858). *Of the Causes Nature and Treatment of Palsy and Apoplexy.* Philadelphia, Lea and Blankard.
Cotzias, G.C., Van Woert, M.H., and Schiffer, L.M. (1967). Aromatic amino-acids and modification of parkinsonism. *N. Engl. J. Med.* 276:374–379.
Cotzias, G.C., Papavasiliou, P.S., and Gillene, R. (1968). Experimental treatment of parkinsonism with L-dopa. *Neurology* 18:276–277.
Cotzias, G.C. (1968). L-dopa for parkinsonism. *New Engl. J. of Med.* 278:630.
Courier & Evening Gazette (1794). *"Crimping"* [Report about Incarceration of Lunatic]. 14 October.
Courier & Evening Gazette (1794). *Letter to Editor from Lunatic's Mother.* 23 October.
Critchley, M. (1929). Arteriosclerotic parkinsonism. *Brain* 52:23.
Crosfield, Robert T. (1797). *Remarks on The Scurvy.* London, The Author.
Cunningham, G.H. (1927). *London. Being a Comprehensive Survey of the History, Tradition, & Historical Associations of Buildings & Monuments etc.* London & Toronto, J. Dent & Sons.
Day, Thomas (1783–89). *The History of Sandford & Merton.* London, J. Stockdale.
Defoe, Daniel (1706). *Jure Divino: A Satyr. In Twelve Books. By the Author of the True-Born Englishman.* London, N.P.
Delay, J., Deniker, P., and Hare, J.M. (1952). Utilisation en therapeutique psychiatrique d'une phenothiazine d'action centrale elective. *Am. Med. Psychol.* 110:112–117.
Denny-Brown, D. (1962). *The Basal Ganglia and Their Relation to Movement Disorders.* London, Oxford, pp. 37–49.
Dictionary of National Biography (1909). London, Smith, Elder.
Duckworth, Dyce (1879). *On a Case of Unilateral Paralysis Agitans* Lancet, Vol. II, pp. 341–2.
Dunn E.H. (1917). Primary and secondary findings in a series of attempts to transplant cerebral cortex in albino rat. *Comp. Neurol.* 27:565–582.
Duvoisin, R. (1987). History of parkinsonism. *Pharmac. Ther.* 32:1–17.

Duvoisin, R.C. and Schweitzer, M.D. (1966). Paralysis agitans mortality in England and Wales 1855–1862. *Br. J. Prev. Soc. Med.* 20:27–33.

Earle, K.M. (1966). Introduction to the J. Neurosurg. 24 Part II:247–249.

Eaton, Daniel Isaac (1794). *Politics for the People (Periodical).* London, The Author.

Ebin, J. (1849). Combined Laberal and Ventral pyramidatomy in treatment of paralysis agitans. *Arch. Neurol. Psych.* 62:27–35.

Ehringer, H. and Horneykiewicz, O. (1960). Verteilung von Noradrenalin und Dopamin (3-hydroxytyramin) im Gehirn des Menschen und ihr Verhalten bis Erkrankungen des extrapyramidalen Systems. *Klin. Wschr.* 38:1236–1240.

Elliotson, John (1830). *Clinical Lecture Given at St. Thomas's Hospital, Oct. 11, 1830.* Reprinted in Lancet, Vol. II, pp. 119–123.

Elliotson, J. (1830). *Notes of a Course of Lectures on the Practice of Medicine.* London, S. Highley, pp. 138–139.

Elliotson, J. (1839). *Principles and Practice of Medicine.* London, J. Butler.

Ellis, Sir Henry (1798). *The History and Antiquities of the Parish of St. Leonard Shoreditch and Liberty of Norton Folgate.* London, Lackington.

Erb, W. (1901). Paralysis agitans (parkinsonische krankheit). In: *die Deutsche Klinik* (E. Leydon and F. Klemperer, eds.). Berlin, Ruban and Schwartzenberg, pp. 31-50.

Fagge, C.H. (1886). *The Principles and Practice of Medicine.* Edited by P.H. Pye-Smith. 2 vols. London, J. and A. Churchill.

Fairchild, Thomas (1722). *The City Gardner, Containing the Most Experienced Methods of Cultivating . . . Such Evergreens . . . Shrubs . . . Exotick Plants, Etc . . . As will Thrive Best in the London Gardens.* London, N.P.

Falck, B., Hillarp, N.A., Thieme, G., et al. (1962). Fluorescence of catecholamines and related compounds condensed with formaldehyde. *J. Histochem. Cytochem.* 10:348–354.

Farey, John (1811–17). *General View of the Agriculture and Minerals of Derbyshire.* 3 vols. London, G. & W. Nicol.

Flerkó, B., Szentagothai, J. (1957). Oestrogen sensitive nervous structures in the hypothalamus. *Acta Endocrinol.* 26:121–127.

Fourcroy, A.C. (1804). *A General System of Chemical Knowledge.* London, Cadell & Davies.

George, M. Dorothy (1931). *England in Transition: Life and Labour in the Eighteenth Century.* London, G. Routledge & Sons.

Godwin, William (1793). *An Enquiry Concerning Political Justice and its Influence on General Virtue and Happiness.* London, L. White.

Good, J.M., Gregory, O., and Bosworth, N. (1813). *Pantologia. A New Cyclopaedia.* London, N.P.

Goodwin, A. (1979). *The Friends of Liberty.* London, Hutchinson.

Gowers, W.R. (1888). *A Manual of Diseases of the Nervous System, 2nd edition.* Philadelphia, Blakiston, pp. 1002–1003.

Gowers, William (1886–88). *Manual of the Diseases of the Nervous System.* 2 vols. London, J&A Churchill.

Grasset, J. and Rauzier, G. (1894). *Traité Pratique des Maladies du Système Nerveux,* 4th ed. Montpellier, C. Coulet.

Graves, R.J. (1842). *A System of Clinical Medicine.* Dublin, Fannin, p. 714.

Gregory, George (1828). *Elements of the Theory and Practice of Physic Designed for the Use of Students,* 3rd ed. London, Burges and Hill.

Hadden, W.B. (1881). *Case of paralysis agitans.* British Medical Journal, Vol. I, pp. 1936–37).

Hall, Marshall (1838). *Lectures on the theory and practice of medicine.* Lancet, Vol. II, pp. 33–42.

Hall, M. (1841). *On the Diseases and Derangements of the Nervous System.* London, Balliere, p. 320.

Hammond, W.A. (1871). *A Treatise on Diseases of the Nervous System.* New York, Appleton, pp. 706–710.

Handfield, Jones (c. 1864). *Clinical Observations on Functional Nervous Diseases.* London, J. Churchill.

Hardy, Thomas (1832). *Memoir of T.H., Founder of . . . The London Corresponding Society . . . Written by Himself.* London, N.P.

Harrison, G.B. (1934). *Introducing Shakespeare.* Harmondsworth, Penguin Books.

Hart, T.S. (1904). Paralysis agitans: some clinical observations based on the study of 219 cases seen at the clinic of Professor M. Allen Starr. *J. Nerv. Ment. Dis.* 31:177–178.

Hassler, R. (1938). Zur Pathologie des Paralysis agitans und des postenzaptalitischen Parkinsonismus. *J. Psychol. Neurol.* 48:387–476.

Haygarth, John (1805). *A Clinical History of Diseases. Pt. I: 1. A Clinical History of Acute Rheumatism. 2. A Clinical History of the Nodosity of Joints.* London, Cadell & Davies.

Henry, William (1801). *An Epitome of Chemistry.* London, B. Johnson.

Henry, William (1810). *Elements of Experimental Chemistry.* London, Johnson.

Hesselink, J.M.K. (1938). Some historical aspects of Parkinson's disease. The progression of insights in the period 1817–1868. *Janus* 70 (3-4):263–279.

Hoehn, M.M. and Yahr, M.D. (1967). Parkinsonism: onset, progression and mortality. *Neurology* 17:427–442.

Hollingshead, John (1895). *My Lifetime.* London, Sampson Low, Marston & Co.

Hope, J.A. (1982). *For the cause of truth: Radicalism in London 1796–1821.* Oxford, Clarendon Press.

Horsley, V. and Clarke, R.H. (1908). The structure and function of the cerebellum examined by a new method. *Brain* 31:45–124.

Hunt, J.R. (1917). Progressive atrophy of the globus pallidus. *Brain* 11:58–148.

Jackson, H.J. (1899). On certain relations of the cerebrum and cerebellum (on rigidity of hemiplegia and of paralysis agitans. *Brain* 22:621–630.

Jones, Charles Handfield (1864). *Clinical Observations on Functional Nervous Diseases.* London, J. Churchill.

Kinglake, Robert (1804). *A Dissertation on Gout.* London, J. Murray.

Kinglake, Robert (1807). *Strictures on Mr. Parkinson's Observations on the Nature & Cure of Gout.* Taunton, Poole.

Kinglake, Robert (1807a). *Additional Cases of Gout in Farther Proof of the Efficacy of the Cooling Treatment.* Taunton, Poole.

Kippis, Andrew (1778–93). *Biographia Britannica. The Second Edition, With Corrections, Enlargements, and the Addition of New Lives.* London, L. Bathurst.

Kirkland, Thomas (1792). *Commentary on Apoplectic and Paralytic Affections.* London, W. Dawson.

Knoll, J. (1983). Deprenyl (selegiline): the history of its development and pharmacological action. *Acta Neurol. Scand.* 95:57–80.

Lancet (1888). *Editorial on Parkinsons Disease.* Vol. II, p. 130.

Lancet (1893a). *Editorial on Borate of Soda in Paralysis Agitans.* Vol. I, p. 377.
Lancet (1893). *Abstract of paper by von Ketscher.* Vol. I, pp. 669–70.
Langston, J.W., Ballard, P.A., Tetrud, J.W. and Trevin, J. (1983). Chronic parkinsonian in human due to a product of meperidine-analog synthesis. *Science* 219: 979–980.
Langston, J.W. (1987). From behavioural models to molecular mechanisms. MPTP as a tool to study Parkinson's Disease. In: *Parkinson's Disease: Clinical and Experimental Advances* (F. Clifford Rose, ed.). London, John Libbey, pp. 137–148.
LeGres Clark, W.E. (1940). Neuronal differentiation in implanted foetal cortical tissue. *J. Neurol. Psychiatry* 3:263–284.
Lewy, F.H. (1914). Zur Patholognischen Anatomie der Paralysis agitans. *Dr. Z. Newtterk.* 50:50–55.
Leigh, Charles (1700). *Natural History of Cheshire, Lancashire, and the Peak of Derbyshire.* Oxford, The Author.
L'Hermitte, J. and Cornil, L. (1921). Recherches anatomique sur la Maladie de Parkinson. *Revue neurol.* 28:587–592.
Lindvall, O., Dunnett, S.B., Brundin, P., and Bjorklund, A. (1987). Transplantation of catecholamine-producing cells to the basal ganglia in Parkinson's disease: experimental and clinical studies. In: *Parkinson's Disease: Clinical and Experimental Advances* (F. Clifford Rose, ed.). London, John Libbey, pp. 189–206.
Loudon, I. (1987). *Medical Care and the General Practitioner, 1750–1850.* Oxford, Clarendon Press.
Maclachlan, Daniel (1863). *A Practical Treatise on the Diseases and Infirmities of Advanced Life.* London, J. Churchill.
Madrago, I., Drucker-Colin, R., Diaz, V., Martinez-Mata, J., Torres, C., and Becevril, J.J. (1987). Open microsurgical autograft of adrenal medulla to the right caudate nucleus in two patients with intractable Parkinson's disease. *N. Engl. J. Med.* 316:831–834.
Mantell, Gideon A. (1822). *The Fossils of the South Downs; or, Illustrations of the Geology of Sussex.* London, L. Relf.
Mantell, Gideon A. (1850). *A Pictorial Atlas of Fossil Remains.* London, H. Bohn.
Marcet, Jane (1806). *Conversations on Chemistry.* London, Longman, Hurst, Rees & Orme.
Marshall, Andrew (1815). *The Morbid Anatomy of the Brain.* London, Longman, Hurst, Rees & Orme.
Martin, J.P. (1967). The globus pallidus in post-encephalitic parkinsonism. *J. Neurol. Sci.* 2:344–347.
Matey, M. (1757–1784). *Medical Observations and Inquiries.* 3 vols. London, Johnston & Cadell; Society of Physicians in London.
McMenemey, W.H. (1955). *James Parkinson 1755–1824: A Biographical Essay* in Critchley, M. (ed.), *James Parkinson 1755–1824.* London, MacMillan, pp. 1–144.
Medico-Chirurgical Review (1817). *Review of Parkinson's "Essay on The Shaking Palsy".* Vol. 4, pp. 401–08.
Miller, J.S. (1821). *A Natural History of The Crinoidea.* Bristol, N.P.
The Monthly Gazette of Health (1817). Vol. 2, p. 567.
Morning Chronicle (1810). *"Imprisonment in a Madhouse".* Report of Daintree Trial, 31st October 1810.
Morris, A.D. (1957). *Hoxton Square and the Hoxton Academies.* London, The Author.

Morris, A.D. (1958). *The Hoxton Madhouses*. Cambridge, The Author.
Morton, John (1712). *Natural History of Northamptonshire*. London, R. Knaplock & R. Wilkin.
Moyer, H.N. (1911). A new diagnostic sign in paralysis agitants: The cog-wheel resistance of the extremities. *J. Amer. Med. Assoc.* 57:2125.
Muller, J. (1842). *Elements of Physiology*, translated by Wm. Baly. London, Taylor and Walton, Vol. II, p. 88.
Murchison, C. and Cayley, W. (1871). *A Case of Paralysis Agitans*. Trans. Path. Soc. Lond., Vol. 22, pp. 24–27.
Nicholson, William (1796). *First Principles of Chemistry*. London, G.G.J. & J. Robinson.
Nicholson, William (1809). *The British Encyclopaedia, or, Dictionary of the Arts & Sciences*. London, Longman, Hurst, Rees, & Orme.
Nicholson's Journal (1811). *Report of Proceedings of Geological Society*. Vol. 32, pp. 314–5.
Nicholson's Journal (1813). *Report of Proceedings of Geological Society*. Vol. 34, pp. 418–9.
Oldenstein, L. (1867). *Sur la Paralysie Agitante et la Sclerose en Plazques Generalisees*. Mortimer, Paris.
The Oracle (1799). *Report of 9th December 1799*.
Oliver, L.O. (1952). Surgery in Parkinson's disease: complete section of the lateral column of the spinal cord for tumor. *Lancet* 1:847–848.
Osler, W. (1892). *The Principles and Practice of Medicine*. New York, Appleton, pp. 926–927.
Paine, Thomas (1791). *Rights of Man: Being an Answer to Mr. Burke's Attack on the French Revolution [Part the First]*. London, J. Johnson.
Paine, Thomas (1792). *Rights of Man. [Part the Second]*. London, J.S. Jordan.
Parkes, Samuel (1806). *A Chemical Catechism*. London, The Author.
Parkinson, James (1780). *Observations on Dr. H. Smith's Philosophy of Physic*. London, The Author.
Parkinson, James (1789). *"Some Account of the Effects of Lightning"*. Memoirs of the Medical Society of London. Vol. 2, pp. 193;493–503.
Parkinson, James (1793?). *Pearls Cast Before Swine . . . Scraped Together by Old Hubert*. London, D.I. Eaton.
Parkinson, James (1793?a). *The Budget of the People, Collected by Old Hubert*. London, N.P.
Parkinson, James (1793?b). *The Village Association, or the Politics of Edley*. London, N.P.
Parkinson, James (1793). *Address to the Hon. Edmund Burke: From the Swinish Multitude. By Old Hubert*. London, J. Ridgeway.
Parkinson, James (1793a). *Knaves-Acre Association. Resolutions Adopted at a Meeting of Placemen, etc. by Old Hubert*. London, N.P.
Parkinson, James (1794?). *Mast and Acorns, The Food of the Swinish Multitude. By Old Hubert*. London, N.P.
Parkinson, James (1794). *Revolutions Without Bloodshed; or, Reformation Preferable to Revolt*. London, N.P.
Parkinson, James (1794a). *A Sketch of the Most Memorable Events in the History of England from the Landing of Julius Caesar, to the Reign of William the Conquerer. By Old Hubert*. Published in: *Politics for the People (Periodical)*, edited by D.I. Eaton. London, The Author.

Parkinson, James (1795?). *A Sketch, by Old Hubert Whilst the Honest Poor are Wanting Bread*. London, J. Burks.

Parkinson, James (1975?a). *An Account of Some Peculiar Manners & Customs of the People of Bull-land, or the Island of Contradictions by Old Hubert*. London, N.P.

Parkinson, James (1795?b). *The Soldier's Tale, Extracted from the Village Association: With Two or Three Words of Advice, by Old Hubert*. London, N.P.

Parkinson, James (1795). *A Letter from Mr. Parkinson Containing an Account of His Examination Before the Privy Council*. London, N.P.

Parkinson, James (1796?). *A Vindications of the London Corresponding Society*. London, N.P.

Parkinson, James (1800). *The Hospital Pupil; or, an Essay Intended to Facilitate the Study of Medicine & Surgery. In Four Letters*. London, H.D. Symonds.

Parkinson, James (1800a). *The Chemical Pocket-Book; or, Memoranda Chemica; Arranged in a Compendia of Chemistry*. London, N.P.

Parkinson, James (1801). *Medical Admonitions to Families, Respecting the Preservation of Health, & The Treatment of the Sick*. London, H.D. Symonds.

Parkinson, James (1802). *The Way to Health . . . Extracted from the Villagers Friend & Physician*. London, H.D. Symonds.

Parkinson, James (1802a). *Hints for the Improvement of Trusses*. London, H.D. Symonds.

Parkinson, James (1803). *The Town & Country Friend & Physician . . . With Cursory Observations on the Treatment of Children*. Philadelphia, J Humphreys.

Parkinson, James (1804). *The Villagers Friend & Physician*, 2nd, Ed. London, C Whittingham.

Parkinson, James (1804–11). *Organic Remains of a Former World*. London, J Robson.

Parkinson, James (1805). *Observations on the Nature & Cure of Gout; on Nodes of the Joints; & on the Influence of Certain Articles of Diet in Gout, Rheumatism, & Gravel*. London, H.D. Symonds.

Parkinson, James (1807). *Remarks on Mr. Whitbread's Plan for the Education of the Poor*. London, H.D. Symonds.

Parkinson, James (1807a). *Letter* Gentleman's Magazine, Vol. 77 (2) p. 818.

Parkinson, James (1807b). *Letter Circulated to Parishioners in the Parish of St. Leonards Concerning the Introduction of a Register of Poor Children Seeking Employment*.

Parkinson, James (1808). *Dangerous Sports. A Tale Addressed to Children*. London, H.D. Symonds.

Parkinson, James (1809). Articles on "Geology," "Oryctology," "Rocks," "Shells," "Conchology" in: *The Britisch Encyclopedia; or, Dictionary of Arts and Sciences*, edited by W. Nicholson. 6 vols. London, Longman, Hurst, Rees & Orme.

Parkinson, James (1811). *"Observations on Some of the Strata in the Neighbourhood of London, and on the Fossil Remains Contained in Them"*. Transactions of the Geological Society, Vol. 1, p. 324.

Parkinson, James and John (1811). *"A Case of Trismus"*. Transactions of the Medical & Chirurgical Society, Vol. 2, p. 291.

Parkinson, James (1811a). *Observations on the Act for Regulating Madhouses*. London, N.P.

Parkinson, James (1814). *"To the Editors of the London Medical Repository."* London Medical Repository, Vol. 1, pp. 289–292.

Parkinson, James (1814a). *"Some Observations on the Specimens of Hippurites Presented to the Society by the Hon. H.G. Bennett."* Transactions of the Geological Society, Vol. 2, p. 277.

Parkinson, James and John (1816). *"A Case of Diseased Vermiform Appendix".* Transactions of the Medical & Chirurgical Society, Vol. 3, p. 57.
Parkinson, James (1817). *An Essay on the Shaking Palsy.* London, Sherwood, Neely & Jones.
Parkinson, James (1819). *Observations on the Necessity for Parochial Fever Wards.* London, N.P.
Parkinson, James (1819a). *"Remarks on the Fossils Collected by Mr. Phillips Near Dover and Folkestone".* Transactions of the Geological Society, Vol. 5, p. 52.
Parkinson, James (1822). *Outlines of Oryctology.* London, The Author.
Parkinson, James (1824). *Obituary Notice.* Gentleman's Magazine, Vol. 94(2), p. 644.
Parkinson, James and John (1824). *"On the Treatment of Infectious or Typhoid Fever."* London Medical Repository, Vol. 1, p. 197.
Parkinson, J.W.K. (1833). *Hunterian Reminiscences . . . a course of Lectures on the Principles & Practice of Surgery.* London, Sherwood, Gilbert & Piper.
Pearson, Hesketh (1942). *A Life of Shakespeare.* Harmondsworth, Penguin Books.
Pennant, Thomas (1793). *Some Account of London.* London, R. Faulder.
Phillips, William (1819). *Remarks on the Chalk Cliffs in the Neighbourhood of Dover, and on the Blue Marl Covering the Green Sand Near Folkestone.* Transactions of the Geological Society, Vol. 5, p. 16.
Pletscher, A., Shore, P.A., and Brodie, B.B. (1955). Serotonin release as a possible mechanism of reserpine action. *Science* 122:374.
Plot, Robert (1677). *Natural History of Oxfordshire.* Oxford, At The Theatre.
Plot, Robert (1685). *Natural History of Staffordshire.* Oxford, At The Theatre.
Porter, C.C., Watson, L.S., Titus, D.C., Tolaro, J.A., and Byer, S.S. (1962). Inhibition of dopa carboxylase by the hydrazine analog of methyl dopa. *Biochem. Pharmacol.* 11:1067–1072.
Price, Richard (1789). *A Discourse on the Love of our Country, Delivered on Nov. 4, 1789.* London, G. Stafford.
Prus, J. (1898). Über die Leitungsbahnen und Pathogenese der Rindenepiecpsie. 858–863.
Prus, J. (1899). Über del bei electrischer Reizung des Corpus striatum und des Thalamus opticus auf Tretenden Erschlinungen. Wien, Klim. Wschr. 48.
Ralston, H.J. and Liber, B. (1953). The question of tonus in skeletal muscle. *Am. J. Physical Med.* 32:85–92.
Ranson, S.W. (1909). Transplantation of the spinal ganglion into the brain. *Q Bull Northwest Univ Med School* 11:176–178.
Rees, Abraham (1819). *The Cyclopedia; or Universal Dictionary of Arts, Sciences, and Literature.* 39 vols. London, Longman, Hurst, Rees & Orme.
Refkin, A. (1987). Extrapyramidal side effects: A historic perspective. *J. Clin. Psychiatry* 48:9 (Suppl) 3–6.
Reynolds, J. Russell (1866–79). *A System of Medicine.* London, N.P.
Ring, John (1816). *An Answer to Dr. Kinglake: Showing the Danger of his Cooling Treatment of the Gout.* London, N.P.
Robinson, William (1842–43). *The History and Antiquities of the Parish of Hackney, in the County of Middlesex.* 2 vols. London J.B. Nichols & Son.
Romberg, M.H. (1851). Lehrbuch der Nervenkrankheiten des Menschen, 2nd Ed. Berlin, Duncker.
Romberg, M.H. (1851). Lehrbuch der Nervenkrankheiten, 2nd ed. Berlin, Duncker, p. 371.

Rose, T.G. and Haskef, L. (1987). Parkinson's disease: Further steps forward. *Gerontology* 33:369–373.

Rowntree, L.G. (1912). *James Parkinson.* Bulletin of the Johns Hopkins Hospital. Vol. 23, pp. 33–45.

Sachs, O.W. (1971). Parkinsonism—a so-called New Disease. *Brit. Med. J.* ii: 111.

Salty Kow, S. (1905). Versuche über Gehirnreplantation zugleich ein Beitragzur Kenutruss reactiver Vorgänge an den zellingen 'Gehirneelem. *Arch. Psychiat. Nervenk* 40:329–388.

Sanders, W.R. (1879). Paralysis agitans. In: *A System of Medicine,* 2nd ed. (J.R. Reynolds, ed.). London, Macmillan, Vol. 5, p. 1056.

Schaltenbrand, G. (1975). Personal observations on the development of stereotoxy. *Confinia. Neurol.* 37:410–416.

Schiller, F. (1986). Parkinsonian rigidity: The first hundred-and-one years, 1817–1918. *Hist. Phil. LIfe. Sci.* 8:221–236.

Schlotheim, E.F. (1820–23). *Die Petrefactenkunde Auf Ihrem Jetzigem Standpunkte Durch Die Beschreibung Seiner Sammlung Versteinerter und Fossiler Überreste des Thier—und Pflanzenreichs der Vorwelt.* Gotha, Becker.

Schwab, R.S., England, A.C., Poskanzer, D.C., and Young, R.R. (1969). Amantadine in the treatment of Parkinson's disease. *J.A.M.A.* 208:1168–1170.

Selden, John (1689). *Table-Talk: Being the Discourses of John Selden, Esq.; or His Sence of Various Matters of Weight and High Consequence . . . [ed. R. Milward].* London, Tonson & Churchill.

Serres, E.R.A. (1824–26). *Anatomie Comparee du Cerveau dans les Quatre Classes des Animaux Vertebres.* Paris, Gabon.

Sinkler, W. (1886). Diseases of the Nervous System. In: *A System of Practical Medicine by American Authors* (W. Pepper, ed.). Philadelphia, Lea, Vol. 5, pp. 42–45, 433–438.

Sladek, J.R. and Gash, D.M. (1988). Nerve-cell grafting in Parkinson's disease. *J. Neurosurg.* 68:337–351.

Smith, Hugh (1780). *Philosophical Inquiries Into the Laws of Animal Life.* London, L. Davis.

Smith, John (1795). *Assassination of the King! The Conspirators Exposed. To Which is Added a Letter From Mr. Parkinson Containing the Particulars of His Examination Before the Privy Council etc.* London, N.P.

Smith, William (1817). *Stratigraphical System of Organized Fossils.* London, E. Williams.

Sowerby, James (1812–46). *The Mineral Conchology of Great Britain. Continued [Vols. 5–7] by J. De C. Sowerby.* London, B. Meredithy.

Spatz, H. (1922). Uber Beziehungen wischen der substantia nigra des Mittelhunfusses und dem Globus pallidus des Linsen Kerns. *Verh. anat. Ges. (Jena)* 31: 159–180.

Spiegel, E.A. and Wycis, H.T. (1950). Principes et applications de la stereoencephalotomie. *Acta neurochir.* 1/2–3.

The Statesman (1810). *"Middlesex Sessions. Conspiracy." Report, 31st Oct.*

The Statesman (1810a). *Letter Describing Maltreatment of a Lunatic. 19th Sept.*

The Statesman (1810b). *Editorial on the Daintree Case. 31st Oct.*

The Statesman (1810c). *Letter Dated 31st Oct. to the Editor Containing Attack on Parkinson. 14th Nov.*

Sunderland, Septimus P. (1915). *Old London Spas, Baths, and Wells.* London J. Bale, Sons & Danielson.

Thompson, W.G. (1890). Successful brain grafting. *N.Y. Med. J* 51:701.

Thompsons Annals of Philosophy (1809). *Report of Preceedings of Geological Society,* Vol. 5, p. 452.

The Times (1793). *Advertisement in Times of 17th Dec.*

The Times (1810). *Report of Daintree Trial, 31st Oct.*

Toulmouche, A. (1833). Observations de quelques functions involuntaires des appareils de la locomotion et de la prehension. *Memoires de l'Academie de Medicine.* 2:368–398.

Trétiakoff, C. (1919). Contribution a l'etude de l'anatomie pathologique du Louis niger de Soemmering avec quelques reductions relative a la pathogenie des troubles du tomes musculaire de la maladie de Parkinson. Thesis, University of Paris.

Trevelyan, G.M. (1949). *Illustrated English Social History.* 4 vols. London, Longmans & Co.

Trousseau, A. (1859). *Des Charecs Union Medicale 2nd Series* 1:246–249.

Trousseau, A. (1862). *Clinique medicale de L'Hotel-Dieu, Vol 2,* pp. 64–180.

Trousseau, A. (1867–1871). *Lectures on Clinical Medicine.* Delivered at the Hotel Dieu, 1867–1871, translated by Victor Bazue. Philadelphia, Lindsay and Blakiston, pp. 441–444.

Trousseau, A. (1868). *Lectures on Clinical Medicine.* trans. P.U. Bazáre. London, New Sydenham Society.

Vincent, Thomas (1667). *God's Terrible Voice in the City.* London, Calvert & Sprint.

Vogt, M. (1954). *J. Physiol. (London).* 123:451–481.

Vogt, O. and Vogt, C. (1920). Zur Lehre der Erkrawkunger der striaren systeme. *J. Psychol. Neurol. Lpz. Suppl.* 25:627–846.

Von Economo, C. (1917). Encephalitis lethargica. *Wien Klin. Wschr.* 30:581–585.

Von Economo, C. (1931). *Encephalitis lethargica,* translated by K.O. Newman. London, Oxford University Press.

Von Oppolzer, J. (1861). Fall von paralysis agitans. *Wien Med. Wehschr. Spitalz.* 265:249–251.

Walker, E.D. (1952). Cerebral pedunculotomy for the relief of involuntary movements: Parkinsonian tumor. *J. Nerv. Oncert. Dis.* 116:266–755.

Watson, Richard (1781–87). *Chemical Essays.* London, T. Evans.

Watson, Sir Thomas (1843). *Lectures on the Principles and Practice of Physic.* 2 vols. London, J.W. Parker.

Wheatley, H.B. (1891). *London Past and Present: its History, Associations, and Traditions.* 3 vols. London, J. Murray.

Wilkins, R.H. and Brodie, I.A. (1968). Encephalitis lethargica. *Arch. Neurol.* 18: 324–328.

Wilson, A.A.K. (1925). The Croonian Lectures in some disorders of motility and muscle tone, with special reference to the corpus striatum. Lecture 1. *Lancet* ii:54–62.

Index

A

Accidental injury, prevention of, 59
Adrenal medullary transplants, 187
Agriculture, and population growth, 5
Air, vital vs. azotic, 56
Akinesia, 179
Althaus, Julius, 143, 148
Alzheimer, A., 180
Amantadine, 184
American Revolution, 3
Animal substances, composition of, 91–92
Anton, 180
Appendicitis, 72–73
Apprentices
  regulations for, 83–84
  victimization of, 79–81
Armstrong, Dr. John, vii
Artificial respiration, 70
Association of Apothecaries and Surgeon–Apothecaries of England and Wales, 22

B

Babington, William, 89, 123
Bamberger, 143, 146
Bampton, A. H., 146
Banks, Sir Joseph, 129
Barbeau, A., 183, 184
Basal ganglia, 180–181
Baxter, John, 32, 34
Beaumont, John, 114
Bechet, E., 176
Beck, John, 53
Beddoes, Thomas, 90
Bedlam, 2
Beevor, Dr. C. E., 146
Benedikt, M., 178
Berger, 179
Bergman, T. O., 93
Bethlem, 97
Birkmeyer, W., 183
Black, Dr. Joseph, 90
Bleeding, 57, 64
Blizard, Sir William, 16, 72, 74
Blocq, P., 180
Blood, composition of, 92
Bonney, John Augustus, 32, 34
Bostock, J, 93
Bournon, Louis de, 123
Bradykinesia, 179
Briellat, Thomas, 30, 31
Brissaud, E., 176, 180
Bristow, Dr. John S., 144
British Medical Journal, 146
British Mineralogical Society, 123
Brougham, Henry, 79
Bruxton, Dr. Isaac, 82
Buchan, Dr., 55
Bucy, P. C., 185
Burdett, Sir Francis, 36, 79
Burke, Edmund, 16, 25, 34
Burrows, Mrs. Ester, 96, 97, 102
Burrows' House, *see* Holly House
Buzzard, Farquàr, 181
Buzzard, Thomas, 131, 144, 146

C
Cancer, treatment of, 59–60
Carlsson, A., 182, 183
Cayley, Dr. W., 145
Chaptal, J. A., 88, 89
Charcot, Jean-Martin, 23, 141–143, 146, 147, 176, 177, 178, 179, 182
Chawner, Rev. George, 101
Chawner v. Warburton, 101, 113
Chemical Essays (Watson), 89, 115
Chemical Pocket Book (Parkinson), 19, 88–95
Children, care and treatment of, 59, 61
Chlorpromazine, 182
Chorea, 141, 176
Clift, William, 15
Cline, Henry, 16
Cogan, Dr. T., 68
Cohn, B., 177
Collins, J., 179
Committee of Correspondence, 17, 43, 46, 51–52
Committee of "Secrecy," 46, 51–52
Constitutional Club, 24, 33
Cooke, J., 176
Cooper, Astley, 16
Cooper, Irving, 186
Copland, J., 177
Cotzias, G. C., 184
Crosfield, Dr. Robert Thomas, 17, 35, 38–48
Cullen, 152, 163
Cuvier, Baron Georges Léopold, 120

D
Daintree, Mary, lawsuit involving, 20, 101–113
Dale, Samuel, 114
Davis, George, 98–100
Davy, Sir Humphry, 89, 90, 93, 94, 129
Decarboxylase inhibitors, 184
Dejerine, J. J., 181
Delay, J., 182
Denny-Brown, D., 180
Dissection, of corpses, 137–138
Disseminated sclerosis, 141
Dopamine, 182–184
Dopamine agonists, 185
Drowning, resuscitation from, 15, 57, 68–71
Drugs, as cause of parkinsonism, 182
Drunkenness, 3
Duckworth, Dr. Dyce, 145
Dumeril, A. H. A., 120

E
Earle, K. M., 180
Eaton, Daniel Isaac, 17, 28
Education, Whitbread's plan for, 18
Edwards, F. E., 121
Electrical treatment, 57
Electricity, chemical effects of, 93
Elliotson, Dr. John, 138, 176
Encephalitis lethargica, 180
  parkinsonism caused by, 181–182
Epilepsy, 59
Erb, W., 179
Erskine, Thomas
  and trial of Hardy, 32, 33
  and trial of Thelwall, 34
Essay on the Shaking Palsy, An (Parkinson), vii, 23, 83, 131–148
  early reaction to, 23, 138
  medical notice of, 138–139
  text of, 151–175
Evans, Thomas, 35, 45–47
"Extrapyramidal system," 181
Eyre, Sir James, 32

F
Fevers, Parkinson on, 55, 75
Fever wards, 81, 82, 84–87
Flood, Noah's, and fossil study, 117, 118, 121
Foetal grafting, 187
Fossils, interest in, 19, 114–123
Fourcroy, A. F., 88, 90
Fox, Charles James, 25, 28
France, war with England, 29, 34, 77, 78, 120

Francis, Philip, 28, 33
French Revolution, 3, 16, 24, 26, 34, 77, 115
English reaction to, 25, 36
Frost, John, 28
Fumigation (for prevention of disease), 54–55

G
Galen, 141, 152, 160, 176
Gaubius, 160, 161
Genesis, Book of, and account of creation, 121
Geological Society of London, 20, 123–124
Geology, *see also* Palaeontology
definition of, 125–126
early science of, 114–115
of London, 124–125
George III (King), 5, 23
plot to assassinate (Pop-Gun Plot), 17, 39–53
George IV (King), 77, 78
Gerrald, Joseph, 28, 31
Glial cells, 179
Gout, 19, 62–68, 83
Gowers, Dr. William, 131, 142, 144, 146, 148, 178, 179
Grasset, J., 143
Graves, R. J., 177
Gubler, 142
Gum, lancing of, 56

H
Habeus Corpus, suspension of, 32, 35, 41, 47
Hackney, 115, 119
Hadden, Dr., 146
Hall, Sir James, 94
Hall, Dr. Marshall, 139, 146, 176
Hammond, W. A., 179
Hardy, Thomas
annual dinner honoring, 35, 37
and London Corresponding Society, 26, 27, 49
trial of, 32–34
Harkness, John, 72
Harris, Wilfred, 181
Hart, T. S., 179
Harveian Society, viii
Hassler, R., 180, 181
Hawes, Dr. W., 68
Henry, William, 94
Hernias, 19, 58–59, 66–67
Higgins, George, 17, 38, 39, 41, 44–49
Hodgson, Richard, 31, 32, 49, 51
Holcroft, Thomas, 32, 34
Holly House (Hoxton), 3, 20, 96, 98, 101
lawsuit of patient detained in, 101–112
Home, Sir Everard, 15
Horsley, Victor, 185, 186
Howard, John, 3
Hoxton, 1
exodus from London to, 5–6
overcrowding and slums in, 6, 8
Parkinson family practice in, 11–13
typhus fever in, 75
workhouse in, 3, 7
Hoxton, Liberty of, 77
Hoxton House, 97
Hoxton madhouses, 2–3, 20, 96–113
Hunt, Ramsay, 180
Hunter, John, 15, 20, 114
Hunter, William, 20
Hunterian Society, 22
Hutton, James, 94, 114, 117, 125
Hydrophobia, 59, 73–75
Hyoscine, 182
Hypochondriacs, 58

I
Industrial Revolution, 5
Insanity, *see also* Lunatics
certificate of, 101–102, 112–113
Ireland, planned insurrection in, 35, 45, 46
Irish
and extreme radicalism, 18
poverty of among immigrants, 6

J
Jackson, Hughling, 177, 179
Jenner, Dr. Edward, 20
Jones, Dr. Handfield, 141, 177
Journal of Natural Philosophy, Chemistry, and the Arts, 7, 20, 89
Joyce, Jeremiah, 32, 34
Juncker, Johann, 132, 135, 152, 159

K
Keys, John, 114
Kinglake, Dr. Robert, 19, 64–65
Kirkland, Thomas, 136
Kyd, Stewart, 32, 34

L
Laborers, Irish immigrant, 6
Lamarck, Jean Baptiste, 120
Lancet, 147, 181
Lavoisier, A. L., 88, 90
Lebert, 143, 146
Leeches, 61, 64
Leigh, Charles, 114
Le Maitre, Paul Thomas, 38, 39, 41, 44–49
Lettsom, Dr. J. C., 16, 68, 71
Lewy bodies, 180
Lightning, effects of, 71
Linnaeus, C., 163
Linnean classification, 120, 121
Lister, Martin, 114
Literature, pursuit of, 7
London, City of
  geology of, 124–125
  prosperity of, 5
London Chemical and Philosophical Society, 15
"London clay," 125
London Corresponding Society, 16–18, 78
  agents-provocateurs in, 32, 38
  Committee of Correspondence of, 17, 43, 47, 51–52
  and fear of insurrection, 30, 31
  foundation of, 26–27
  and French Revolution, 27–28
  leaders of, 23
  manifesto of, 27
  meetings of (1793), 30–31
  membership decline in, 28–29
  parliamentary attempts to crush, 35
  and parliamentary reform, 26–28, 36, 37
  and Pop-Gun plot, 38–53
  and Scottish reform movement, 30–31
  "Secret Committee" of, 46, 51–52
  and trials of Thelwall and others (1794), 34
London Electrical Dispensary, 57
London Hospital, 2, 15, 82
London Medical and Physical Journal, 85
London Medical Repository, 16, 75
Louis XVI (King of France), 24, 26, 29
Lovett, John, 32
Loyal Lambeth Association, 32, 33
Lunatics
  and Hoxton madhouses, 2–3, 20, 96–98
  lawsuits over detention of, 101–113
  treatment of, 3, 96–98
Lyell, Charles, 128
Lynam, George, 32

M
Macartney, Dr. James, 120
Maclachlan, David, 140
Macqueen, Robert, 30
Madhouse(s), *see also* Lunatics
  Bedlam, 2
  Burrows, *see* Holly House
  at Hoxton, 2–3, 20, 96–98
  private, 22, 96, 106, 113
  of Sir Jonathan Miles, 22
Malthus, Thomas Robert, 5, 18
Mantell, Dr. Gideon, 15, 19, 122–123
Marcet, Dr. Alexander, 82
Marcet, Jane, 94, 95
Margarot, Maurice, 26, 31
Marinesco, G., 180
Marshall, Dr. Andrew, 16, 74–75
Martin, Purdon, 180

Martin, William, 117, 127
Maty, Dr. M., 166, 167
Medical and Chirurgical Society, 7, 8, 71
Medical Society of London, 7, 8, 16, 22, 71
Medico-Chirurgical Society of London, 22, 124
Medulla oblongata, in Parkinson's disease, 143, 146, 164, 167–168, 176
Meynert, Theodor, 180
Middlesex Hospital, 82
Miles, Sir Jonathan, 22
Military service, and crimps and press gangs, 98–100
Monoamine oxidase (MAO) inhibitors, 185
Moore, Matthew, 32
Morris, Arthur Daniel, v, vii–ix
Motor pathway lesioning, 185
MPTP model, 186
Muir, Thomas, 30
Multiple sclerosis, 176, 177
Murchison, Dr. C., 145
Murchison, Sir Roderick Impey, 129
Muriatic acid test, 90, 119
Murray, Dr. John, 145
Muset, David, 91
Muskens, L. J. J., 179

N

Narabayashi, 186
Nautilus Parkinsoni, 121
Neptunist theory, 94, 125
Neural transplantation, 186–187
Neurotransmitters, mono-amine, 182–183
Nicholson, William, 7, 88, 89, 121, 128
Nona, 181
Noradrenaline, 182, 183

O

Oldenstein, L., 179
"Old Hubert" (Parkinson's pseudonym), 17, 28
Ord, Dr., 146
Organic Remains of a Former World (Parkinson), viii, 19–20, 115–122
Oryctology, definition of, 126; *see also* Palaeontology

P

Paine, Thomas, 16, 25, 27, 28, 30
Palaeobotany, 19, 117
Palaeontology, 19–20
  and origins of geology, 114–115
  textbook on, 115–116
Palmer, Peregrine, 38, 40, 42, 44
Palmer, Thomas Fyshe, 30
Palpitation, 160
Paralysis agitans, *see* Parkinson's disease
Parkes, Samuel, 94, 95
Parkinson, James
  on accidental injuries, 59
  on air, vital vs. azotic, 56
  appearance of before Privy Council, 48–53
  on appendicitis, 72–73
  birth and formative period of, 3, 14–15
  and Board of Trustees of Poor, 79–81
  chemistry, interest in, 19, 88–95, 115
  on children, 59, 61
  and Daintree lawsuit, 101–113
  death of, 13–14, 82–83
  on drunkenness, 3
  on effects of lightning, 71
  on electrical treatment, 57
  on epilepsy, 59
  on exclusive patents, 67–68
  on fevers and fever wards, 75, 81, 82, 84–87
  and fossils, 19, 114–123
  on fumigation, 54–55
  geological interests of, 15, 91, 114–115, 124–125
  on hernias, 19, 58–59, 66–67
  and Hoxton madhouses, 20, 97–117

Parkinson, James (*cont.*)
on hydrophobia, 59, 73–75
on hypochondriacs, 58
interest of A. D. Morris in, viii
involvement in reform movement, 16–17, 77–79
on lancing of gum, 56
literary interests of, 7
on Malthus, 18
medical career of, 16, 20
medical education of, 15
as naturalist, 19
as palaeontologist, 114–130
as Parish Doctor, 81
parish involvement of, 77, 79
and Pop-Gun Plot, 43–44, 47–53, 98
on population growth, 5
on pulmonary consumption, 56
on resuscitation, 15, 57, 68–71
and Sunday School, 18
on teething, 56
testimony of at treason trials, 34, 43–44
on treatment of cancer, 59–60
on trismus, 71–72
and vaccination, 20, 61
and victimization of apprentices, 79–81, 83
on weakness of stomach, 58
on worms, 57

Parkinson, James, writings by or attributed to
Address to the Hon. E. Burke from the Swinish Multitude, 17, 25
"Cases of Hydrophobia," 16
Chemical Pocket Book, 19, 88–95
Dangerous Sports, 18
Essay on the Shaking Palsy, An, vii, 23, 62, 83, 131–148, 151–177
Hints on the Improvement of Trusses, 19, 66–67
Hospital Pupil, The, 14, 90
Medical Admonitions to Families, 18, 54–61
Observations of Dr. Hugh Smith's Philosophy of Physic, 15
Observations on the Act for regulating Madhouses, 19
Observations on the Nature and Cure of Gout, 19, 62–68
"Observations on the Necessity for Parochial Fever Wards," 82
Organic Remains of a Former World, viii, 19–20, 63, 115–122
Outlines of Oryctology, 20, 127
political tracts, 17
"Some Account of the Effects of Lightning," 16, 57
"On the Treatment of the Infectious or Typhoid Fever," 75
Villager's Friend and Physician, The, 18
Way to Health, The, 19

Parkinson, James, writings of with John William Keys Parkinson
"A Case of Diseased Vermiform Appendix," 72–73
"A Case of Trismus," 71–72

Parkinson, James Keys (grandson), 12
Parkinson, John (father), 11, 15, 62–63, 68
Parkinson, John William Keys (son), 11–12, 15, 72, 73, 75, 81, 83, 116
Parkinson, Mary (mother), 11
Parkinson, Mary Dale (wife), 11
Parkinson, Mary Hester (granddaughter), 12
Parkinsonism
drug-induced, 182
post-encephalitic, 181–182
Parkinson's disease, *see also Essay on the Shaking Palsy, An*
akinesia in, 179
amantadine in, 184
causes of, 136, 147, 164
description of
by later writers, 138–148
by Parkinson, 133–136, 155–158, 165–171, 175
differentiation from other disorders, 136, 162–163
and dopamine, 182–184
gait in, 134, 145, 153–154, 160–161
history since Parkinson, 176–187

hope for cure of, 137, 139, 172–174
idiopathic vs. post-encephalitic, 181
incidence of, 147–148, 177
initial description of, 132–135, 151–158
localization of lesion in substantia nigra, 176, 179–180, 183, 184
medulla oblongata in, 143, 146, 164, 167–168, 176, 179–180
MPTP model of, 186
neural transplant in, 186–187
origin of term, 23, 131, 143–144
pathology of, 138, 144, 146, 147, 179–180
and post mortem studies, 136, 137, 140, 143, 145, 147
and "rheumatic affections," 136–137, 169
rigidity in, 142, 178–179
signs and symptoms of, 132–136, 144, 153–155
site of lesion in, 176, 179–180
speech in, 144, 145
spinal cord in, 139, 145, 147
surgical techniques in, 185–187
treatment of, 141–142, 145–146, 182–187
tremor in, 131–132, 136, 141, 152–153, 177–178
without tremor, 142
Parkinson's Table (of symptoms), 54–55
Parliamentary reform
bills for (1780s), 23, 24; (1809), 36, 79
and London Corresponding Society, 26–28, 36, 37
movement for, 16, 78
pressure for, 24
Reform Act (1832), 36, 37
reform societies, 30, 35; *see also* London Corresponding Society
Patents, 66–67
Pearson, George, 89, 90, 93
Petersen, Dr Frederick, 148
Petrification, 117, 126
Phillips, William, 123, 127
Philosophical Magazine, 8
Philosophical Transactions (Royal Society), 89
Pitt, William (the Elder), 23
Pitt, William (the Younger), 16, 23–25, 29, 30, 32–34, 36
Place, Francis, 34
Plot, Robert, 114
Plutonist theory, 94, 125
Poor, Board of Trustees for, 79–82
Pop-Gun Plot, 17, 98
initial events in, 38–39
and London Corresponding Society, 38, 41
and Parkinson, 43–44, 47–53, 98
trials for, 40–53
Population growth, 5, 18
Post-encephalitic parkinsonism, 181–182
Poverty, 6–7, 23, 29
and educational schemes, 18
and treatment of pauper lunatics, 97
Powell, Dr. Richard, 94, 113
Priestley, Dr. Joseph, 28, 33
Privy Council, Parkinson's appearance before, 48–53
Prus, J., 181
Pulermacher chain, 146
Pulmonary consumption, 56
Putnam, 184
Pye-Smith, Dr., 143, 148

R

Rauzier, G., 143
Reflections on the Revolution to France (Burke), 25
Parkinson's response to, 16, 17
Reform Act (1832), 36, 37
Resistance, "cogwheel," 179
Resuscitation (from drowning or suffocation), 15, 57, 68–71
Revolution of 1688, 24
Reynolds, J. Russell, 141, 142, 146
Rheumatic disorders, 136–137, 169
Rhodes, William, 115
Richmond, Duke of, 16
Richter, John, 32, 34
Ridgway, James, 28

Rights of Man, The (Paine), 16, 25, 27, 28
Rigidity, 178–179
Roberts, Dr. E., 82
Roberts, Dr. Frederick, 144
Rock, definition of, 126
Romberg's sign, 178
Royal College of Surgeons, 129–130
Royal Humane Society, 15, 57, 68
Royal Jennerian Society, 20
Royal Society, 129
Ruptures, *see* Hernias

S

Sacaze, Dr., 147
St. Bartholomew's Hospital, 82, 94
St. George's Hospital, 82, 89
St. Leonard's, Shoreditch, viii, 3, 6, 8–9, 11
  Parkinson's involvement in parish of, 18, 77, 79, 81
  and parliamentary reform, 36–37
  Sunday School at, 18
  and victimization of apprentices, 79–80
St. Luke's Hospital, 97
Sanders, Dr. W. R., 148
Sanitation
  fumigation for, 54–55
  in Hoxton, 7
Sauvages, F. B. de, 135, 159, 160, 161
Scelotyrbe festinans, 161
Schwab, R. S., 184
Scientific periodicals, 7–8
Scotland, reform movement in, 30–31, 34
Scott, Sir John (Attorney General), 32, 33, 41
Secret Committee, 46, 51–52
Seditious Meetings Act (1795), 18, 35
See, M. G., 141
Select Committee, and investigation of treatment of lunatics, 97
Selegeline, 185
Shaking palsy, *see* Parkinson's disease
Shells, article on, 126–127
Sheridan, Richard Brinsley, 33
Shoreditch, *see also* St. Leonard's, Shoreditch
  fossils found in, 115, 119
  hospitals at, vii-viii, 2–3
  London Corresponding Society meeting in (1793), 30–31
  medical practice in, 20, 22
  population of, 6, 8
  theater in, 2
Skirving, William, 31
Skoda, 143, 146
Slums, in Hoxton, 6, 8
Smith, Dr. Hugh, 15
Smith, John, 17, 18, 38, 39, 41, 44–49
Smith, William, 124, 125
Smyth, Dr. James Carmichael, 55
Society for Constitutional Information, 16, 24, 28, 31, 32, 34
Society for the Relief of Widows and Orphans of Medical Men, 22
Society of United Irishmen, 45, 46
Southey, Dr. H. H., 82
Spatz, H., 181
Spiegel, E. A., 186
Spinal cord, in Parkinson's disease, 139, 145, 147
Spitalfields, 29
Sports, warnings about, 18–19
Stanhope, Earl, 33
State Trials, *see* Treason trials
Stereotactic techniques, 186
Stomach, weakness of, 58
Strata (geological)
  of London, 124–125
  study of, 20, 115, 128–129
Substantia nigra, in Parkinson's disease, 176, 179–180, 183, 184
Sunday Schools, 3, 18
Sylvius de la Boe, F., 132, 135, 152, 159, 161
Symonds, Henry Delahoy, 28
Symptoms, Parkinson's Table of, 54–55

T

Taxation, 23, 29
Teething, 56

Tessier, M., 147
Thalamic syndrome, 181
Thelwall, John, 16–17, 32, 34, 41, 49
Thompson, Benjamin (Count Rumford), 90
Thomson, Dr. T., 90
Tilloch, Dr. Alexander, 8
Tone, definition of, 178
Tooke, John Horne, 32, 33, 41, 49
  trial of, 34
Toulmouche, A., 176–177
Townsend, Rev. Joseph, 56
Transactions of the Geographical Society, 20
Transplants, neural, 186–187
Treasonable Practices Act (1795), 18, 35
Treason trials
  of Crosfield (1796), 35
  of Hardy, Thelwall, and others (1794), 17, 18
  for Pop-Gun Plot (1795), 17, 41–45
  in Scotland (1793 and 1794), 30, 31
Tremor, 152–153, 177–178
  differentiation of, 131–132, 178
  due to substance abuse or old age, 136
  vs. palpitation, 160
  types of, 141, 158–159
Tremor coactus, 159, 161
Tretiakoff, C., 176, 180
Trismus, 71–72
Trousseau, Dr. Armand, 140, 141, 178
Trusses, advice about, 19, 66–67
Typhus fever, 22, 75
  fresh air in treatment of, 56

U

Upton, Thomas, 38, 40–45, 48–49
Urine, composition of, 92–93

V

Vaccination, 20, 61
van Swieten, G., 132, 135, 141, 159
Virchow, Rudolf, 179
Vogel, W., 163
Vogt, M., 182
von Ketscher, 147
von Economo, 180, 181–182
von Euler, U. S., 182
Von Oppolzer, Johann, 140, 143, 146, 178
Vulpian, E. F., 23, 176, 177, 178

W

Wakefield, Edward, 97
Waldheim, Baron Fischer de, 114
Warburton, Thomas, 97, 101
Ward, Robert, 39
Wardle, Thomas, 32
Watson, Dr. Richard, 15, 89, 115
Watson, Dr. Thomas, 139
Watt, Robert, 33
Werner, A. G., 125, 126
Westminster Hospital, 82
Wharton, John, 31
Whitbread, Samuel, M. P., 18
Whitmore House, 97
Wilson, Kinnier, 176, 181, 186
Winterbottom, Rev. William, 28
Wood, petrification of, 117
Woodhouse, James, 93
Woodward, John, 114
Workhouse(s), 3, 81
  and typhus fever epidemic, 75, 82
Worms, 57
Wycis, H. T., 186

Y

Yelloly, Dr. John, 82
Young, Dr. Thomas, 82